Bob van Hilten
Bart Nuttin

Proceedings of the Medtronic Forum
for Neuroscience and Technology 2005

Bob van Hilten (Editor)

Bart Nuttin (Editor)

Proceedings of the Medtronic Forum for Neuroscience and Neuro-Technology 2005

With 13 Figures and 9 Tables

 Springer

Dr. Bob van Hilten
Department of Neurology
Leiden University Medical Center
PO Box 9600
2300 RC, Leiden
The Netherlands

Prof. Dr. Bart Nuttin
Department of Neurology
Universitaire Ziekenhuizen Leuven
Herestraat 49
3000 Leuven
Belgium

ISBN 978-3-540-32745-5   Springer Berlin Heidelberg New York

Cataloging-in-Publication Data applied for
A catalog record for this book is available from the Library of Congress.

Bibliographic information published by Die Deutsche Bibliothek
Die Deutsche Bibliothek lists this publication in the Deutsche Nationalbibliografie;
detailed bibliographic data is available in the internet at <http://dnb.ddb.de>.

Springer Medizin Verlag.
A member of Springer Science+Business Media
springer.de
© Springer Medizin Verlag Heidelberg 2007

SPIN 11681786

Cover Image: Three scanned neurons visualized using particle emission algorithms

Compliments of **medisee** and the EPFL (École Polytechnique Fédérale de Lausanne), Brain-Mind Institute

Cover Design: deblik, Berlin
Typesetting: Hilger VerlagsService, Heidelberg
Printing and Binding: Stürtz AG, Würzburg

Printed on acid-free paper   18/5135/bk – 5 4 3 2 1

Table of Contents

Part III:
Clinical Research in Interventional Neuroscience

List of Contributors

Albanese, Alberto
Istituto Nazionale Neurologico Carlo Besta,
Università Cattolica del Sacro Cuore,
Via G. Celoria, 11, 20133 Milan MI, Italy

Aminian, Kamiar
Hôpital de Zone Morges, Chemin du Crêt 2,
1110 Morges, Switzerland

Benabid, Alim-Louis
Department of Neurosurgery, Grenoble
University Hospital, Cedex 09, Grenoble, France

Boleaga, Bernardo
Department of Magnetic Resonance Imaging,
Clinica Londres, Mexico City, Mexico

Boon, Paul
Reference Center for Refractory Epilepsy and
Department of Neurology, Ghent University
Hospital, Pintelaan 185, 9000 Ghent, Belgium

Brito, Francisco
Department of Stereotactic and Functional
Neurosurgery, General Hospital of Mexico, and
Department of Medical research in Neuro-
physiology, National Medical Center, Mexico City,
Mexico

Broggi, Giovanni
Department of Neurosurgery, Istituto Nazionale
Neurologico Carlo Besta, Università Cattolica del
Sacro Cuore, Via G. Celoria, 11, 20133 Milan MI,
Italy

Brugières, Pierre
Département des Neurosciences, CHU Henri
Mondor, Créteil, France, and INSERM U421,
Faculté de Médecine, Créteil, France

Buchser, Eric
FMH en Anesthésiologie et en Médecine Intensive, Hôpital de Zone Morges, Chemin du Crêt 2,
1110 Morges, Switzerland

Bugiani, Orso
Department of Neurosurgery and Division of
Neurology, Istituto Nazionale Neurologico
Carlo Besta, Università Cattolica del Sacro Cuore,
Via G. Celoria, 11, 20133 Milan MI, Italy

Bussone, Gennaro
Department of Neurology, Istituto Nazionale
Neurologico Carlo Besta, Università Cattolica del
Sacro Cuore, Via G. Celoria, 11, 20133 Milan MI,
Italy

Caemaert, Jacques
Reference Center for Refractory Epilepsy and
Department of Neurology, Ghent University
Hospital, Pintelaan 185, 9000 Ghent, Belgium

Chabardes, Stéphan
Grenoble University Hospital, Cedex 09,
Grenoble, France

Cosyns, Paul
University Hospital of Antwerpen, Wilrijkstraat 10,
2650 Edegem, Belgium

Cruccu, Giorgio
EFNS Panel on Neuropathic Pain, Department
of Neurological Sciences, La Sapienza University,
Viale dell'Universita 30, 00185, Rome, Italy

De Herdt, Veerle
Reference Center for Refractory Epilepsy and
Department of Neurology, Ghent University
Hospital, Pintelaan 185, 9000 Ghent, Belgium

Durrer, Anne
Hôpital de Zone Morges, Chemin du Crêt 2,
1110 Morges, Switzerland

Ferroli, Paolo
Department of Neurosurgery, Istituto Nazionale
Neurologico Carlo Besta, Università Cattolica del
Sacro Cuore, Via G. Celoria, 11, 20133 Milan MI,
Italy

Fraix, Valérie
Grenoble University Hospital, Cedex 09,
Grenoble, France

Franzini, Angelo
Department of Neurosurgery, Istituto Nazionale
Neurologico Carlo Besta, Università Cattolica del
Sacro Cuore, Via G. Celoria, 11, 20133 Milan MI,
Italy

Gabriëls, Loes
University Hospital of Antwerpen, Wilrijkstraat 10, 2650 Edegem, Belgium

Goadsby, Peter J.
Headache Group, Institute of Neurology,
The National Hospital for Neurology
and Neurosurgery, Queen Square, London
WC1N 3BG, UK

Herholz, Karl
Department of Neurology and Max Planck-
Institute for Neurological Research, Cologne,
Germany

van Hilten, Bob
Department of Neurology, Leiden University
Medical Center, PO Box 9600, 2300 RC, Leiden,
The Netherlands

Kéravel, Yves
Département des Neurosciences, Henri Mondor
Hospital, AP-HP, Paris 12, Créteil, France, and
INSERM U421, Faculté de Médecine, Créteil,
France

Klein, Johannes Christian
Department of Neurology and Max Planck-
Institute for Neurological Research, Cologne,
Germany

Klosterkötter, Joachim
Department of Psychiatry, University of Cologne,
Kerpenerstraße 62, 50937 Cologne, Germany

Koulousakis, Athanasios
Department of Stereotactic and Functional
Neurosurgery, University of Cologne,
Kerpenerstraße 62, 50937 Cologne, Germany

Krack, Paul
Grenoble University Hospital, Cedex 09,
Grenoble, France

Larson, Paul S.
University of California, 505 Parnassus Ave. Rm.
M779, San Francisco, CA 94143-0112, USA

Lefaucheur, Jean-Pascal
Service des Explorations Fonctionnelles,
CHU Henri Mondor, Paris 12, Créteil, France,
and INSERM U421, Faculté de Médecine,
Créteil, France

Lenartz, Doris
Department of Stereotactic and Functional
Neurosurgery, University of Cologne,
Kerpenerstraße 62, 50937 Cologne, Germany

Leone, Massimo
Department of Neurology, Istituto Nazionale
Neurologico Carlo Besta, Università Cattolica del
Sacro Cuore, Via G. Celoria, 11, 20133 Milan MI,
Italy

Levesque, Nadja
University of California, 505 Parnassus Ave. Rm.
M779, San Francisco, CA 94143-0112, USA

van der Linden, Chris
Movement Disorder Center, St. Lucas Hospital
Ghent, Groene Briel 1, 9000 Ghent, Belgium

Linderoth, Bengt
Dept. of Neurosurgery, Karolinska Institutet
and Karolinska University Hospital, Nobels väg 5,
Solna, 171 77 Stockholm, Sweden

Marras, Carlo
Department of Neurosurgery, Istituto Nazionale
Neurologico Carlo Besta, Università Cattolica del
Sacro Cuore, Via G. Celoria, 11, 20133 Milan MI,
Italy

Martin, Alastair J.
University of California, 505 Parnassus Ave. Rm.
M779, San Francisco, CA 94143-0112, USA

Meyers, Jeffrey
University of California, 505 Parnassus Ave. Rm.
M779, San Francisco, CA 94143-0112, USA

Nguyen, Jean-Paul
Département des Neurosciences, Henri Mondor
Hospital, AP-HP, Paris 12, Créteil, France, and
INSERM U421, Faculté de Médecine, Créteil, France

Nuttin, Bart
Department of Neurology, Universitaire Zieken-
huizen Leuven, Herestraat 49, 3000 Leuven,
Belgium

Ostrem, Jill
University of California, 505 Parnassus Ave. Rm.
M779, San Francisco, CA 94143-0112, USA

Palfi, Stephane
Neurosurgery Department, Henri Mondor
Hospital, AP-HP, Paris 12, Créteil, France

Paraschiv-Ionescu, Anisoara
Hôpital de Zone Morges, Chemin du Crêt 2,
1110 Morges, Switzerland

Pollak, Pierre
Grenoble University Hospital, Cedex 09,
Grenoble, France

Porta, Mauro
Department of Neurology: Tourette Syndrome
Center and Movement Disorders Clinic,
Policlinico San Marco, Corso Europa 7, 24040
Zingonia-Osio Sotto(BG), Italy

van Roost, Dirk
Reference Center for Refractory Epilepsy and
Department of Neurology, Ghent University
Hospital, Pintelaan 185, 9000 Ghent, Belgium

Schlaepfer, Thomas E.
Klinik für Psychiatrie und Psychotherapie,
Universitätsklinikum Bonn, Sigmund-Freud-
Straße 25, 53127 Bonn, Germany

Sootsman, W. Keith
University of California, 505 Parnassus Ave. Rm.
M779, San Francisco, CA 94143-0112, USA

Seigneuret, Eric
Grenoble University Hospital, Cedex 09,
Grenoble, France

Starr, Philip A.
Department of Neurological Surgery, University
of California, 505 Parnassus Ave. Rm. M779,
San Francisco, CA 94143-0112, USA

Sturm, Volker
Department of Stereotactic and Functional
Neurosurgery, University of Cologne,
Kerpenerstraße 62, 50937 Köln, Germany

Talke, Pekka
University of California, 505 Parnassus Ave. Rm.
M779, San Francisco, CA 94143-0112, USA

Taylor, Rod
Dept. of Public Health & Epidemiology,
University of Birmingham, Birmingham,
B15 2TT, United Kingdom

Treuer, Harald
Department of Stereotactic and Functional
Neurosurgery, University of Cologne,
Kerpenerstraße 62, 50937 Cologne, Germany

Velasco, Francisco
Department of Stereotactic and Functional
Neurosurgery, General Hospital of Mexico,
and Department of Medical research in
Neurophysiology, National Medical Center,
Mexico City, Mexico

Velasco, Marcos
Department of Stereotactic and Functional
Neurosurgery, General Hospital of Mexico,
and Department of Medical research in
Neurophysiology, National Medical Center,
Mexico City, Mexico

Vonck, Kristl
Reference Center for Refractory Epilepsy and
Department of Neurology, Ghent University
Hospital, Pintelaan 185, 9000 Ghent, Belgium

Part I:

Developments in Neuroscience: Pain and Dystonia

Clinical Features of Dystonia and European Guidelines for Diagnosis and Treatment

A. Albanese

Introduction

Dystonia is characterized by sustained muscle contractions, frequently causing repetitive twisting movements or abnormal postures [1, 2]. Although it is thought to be rare, it is possibly underdiagnosed or misdiagnosed due to the lack of specific clinical criteria.

The prevalence of dystonia is difficult to ascertain. On the basis of the best available prevalence estimates, primary dystonia may be 11.1 per 100,000 of early-onset cases in Ashkenazi Jews from the New York area, 60 per 100,000 of late-onset cases in Northern England, and 300 per 100,000 for late-onset cases in the Italian population over the age of 50 [3].

Primary dystonia and dystonia plus are chronic and often disabling conditions with a wide spectrum mainly in young people. Areas of specific concern include differential diagnosis with other movement disorders, etiological diagnosis, drug treatment, surgical interventions, and genetic counseling.

Methods

A task force appointed by the European Federation of Neurological Societies (EFNS) and by the European Section of the Movement Disorders Society (MDS-ES) met repeatedly to perform a systematic review on the diagnosis and treatment of primary (idiopathic) dystonia and dystonia plus syndromes. Task force members were: A. Albanese (Milan, chairman), M.P. Barnes (Newcastle-upon-Tyne), K.P. Bhatia (London), E. Fernandez (Barcelona), G. Filippini (Milan), T. Gasser (Tubingen), J.K. Krauss (Hanover), A. Newton (Brussels), I. Rektor (Brno), M. Savoiardo (Milan), J. Valls-Solé (Barcelona).

Computerized MEDLINE and EMBASE searches (1966 up to February 2005) were conducted using a combination of textwords and MeSH terms: "dystonia", "blepharospasm", "torticollis", "writer's cramp", "Meige syndrome", "dysphonia" and "sensitivity and specificity" or "diagnosis", and "clinical trial" or "random allocation" or "therapeutic use" limited to human studies. The Cochrane Library and the reference lists of all known primary and review articles were searched for relevant citations. No language restrictions were applied. Studies of diagnosis, diagnostic test, and various treatments for patients suffering from dystonia were considered and rated as level A to C according to the recommendations for EFNS scientific task forces [4]. Where only class-IV evidence was available but consensus could be achieved, we have proposed good practice points.

The results of the literature searches were circulated via e-mail to the task force members for comments. The task force chairman prepared a first draft of the manuscript based on the results of the literature review, data synthesis and comments from the task force members. The draft and the recommendations were discussed during a conference held in Milan on February 11/12, 2005, until consensus was reached within the task force.

Results

Diagnosis

The literature search on the diagnosis of dystonia identified no existing guidelines or systematic reviews. Two consensus agreements [1, 5], two reports of workshops or taskforces [6, 7], 69 primary studies on clinically based diagnosis and 292 primary studies on the diagnostic accuracy of different laboratory tests were found. Dealing with primary clinical studies, there were 6 cohort studies, 23 case-control studies, 3 cross-sectional, and 37 clinical series.

The classification of dystonia is based on three axes: (a)"etiology, (b) age at onset of symptoms, and (c) distribution of body regions affected (❏ Table 1.1). The etiological axis discriminates primary (idiopathic) dystonia, in which dystonia is the only clinical sign without any identifiable exogenous cause or other inherited or degenerative disease, from non-primary forms in which dystonia is usually just one of several clinical signs. Dystonia plus is characterized by dystonia in combination with other movement disorders, for example myoclonus or parkinsonism. Primary dystonia and dystonia plus, whether sporadic or familial, are thought to be of genetic origin in most cases.

The clinical features of dystonia encompass a combination of dystonic movements and postures to create a sustained postural twisting (torsion dystonia). Dystonic postures can precede the occurrence of dystonic movements and in rare cases can persist without appearance of dystonic movements (so called "fixed dystonia") [8]. Sustained dystonic postures may be the presenting feature of torsion dystonia and may remain the only sign for many years before torsional movements become apparent. Dystonia has some spe-

❏ Table 1.1. Classification of dystonia based on three axes

Axis	Definition
By cause (etiology)	- Primary (or idiopathic): dystonia is the only clinical sign and there is no identifiable exogenous cause or other inherited or degenerative disease. Example: DYT-1 dystonia. - Dystonia plus: dystonia is a prominent sign, but is associated with another movement disorder. There is no evidence of neurodegeneration. Example: Myoclonus-dystonia (DYT-11). - Heredo-degenerative: dystonia is a prominent sign, among other neurological features, of a heredo-degenerative disorders. Example: Wilson's disease. - Secondary: dystonia is a symptom of an identified neurological condition, such as a focal brain lesion, exposure to drugs or chemicals. Examples: dystonia due to a brain tumour, off-period dystonia in Parkinson's disease. - Paroxysmal: dystonia occurs in brief episodes with normalcy in between. These disorders are classified as idiopathic (often familial although sporadic cases also occur) and symptomatic due to a variety of causes. Three main forms are known depending on the triggering factor. In paroxysmal kinesigenic dyskinesia (PKD; DYT-9) attacks are induced by sudden movement; in paroxysmal exercise induced dystonia (PED) by exercise such as walking or swimming, and in the non-kinesigenic form (PNKD; DYT-8) by alcohol, coffee, tea, etc. A complicated familial form with PNKD and spasticity (DYT-10) has also been described.
By age at onset	- Early onset (variably defined as 20–30 years): usually starts in a leg or arm and frequently progresses to involve other limbs and the trunk. - Late onset: usually starts in the neck (including the larynx), the cranial muscles or one arm. Tends to remain localized with restricted progression to adjacent muscles.
By distribution	- Focal: single body region (e.g., writer's cramp, blepharospasm) - Segmental: contiguous body regions (e.g., cranial and cervical, cervical and upper limb) - Multifocal: non-contiguous body regions (e.g., upper and lower limb, cranial and upper limb) - Generalized: both legs and at least one other body region (usually one or both arms) - Hemidystonia: half of the body (usually secondary to a structural lesion in the contralateral basal ganglia)

cific features that can be recognized by clinical examination. Speed of contractions of dystonic movements may be slow or rapid, but at the peak of movement it is sustained. The involuntary movement associated with dystonia is often variable within months or years and from one subject to the other. However, during a given period of observation, and in each affected individual, dystonia is distinctively consistent and predictable.

Dystonic Postures

Postures that flex or twist a body part along the main axis are associated with a sensation of rigidity and traction. Dystonic postures are directional and force the involved body region into an abnormal position that is consistently present. In axial dystonia postural abnormalities are often a prominent feature, due to the rare occurrence of dystonic movements in the trunk. Predominantly postural forms of axial dystonia include scoliosis and camptocormia. Usually, pain is not a prominent feature of primary dystonia, except for cervical dystonia [9] and some secondary forms. Dystonic postures, rather than movements, cause pain.

Dystonic Movements

As a rule, dystonic movements have a twisting nature and a directional quality, they are repetitive and patterned, consistent and predictable, and are sustained at their peak. The directional quality is sustained (if only for an instant), and consistency and predictability indicate that the same muscle groups are repeatedly involved. Movements are directional with variable speed. Dystonic neck movements have a directional preponderance, forcing the head to assume an abnormal position (e.g., horizontal rotation or lateral tilt), if only for a moment. Similarly, other focal forms of dystonia result in consistent directional or posture assuming movements (e.g., ulnar deviation, plantar flexion, vocal fold adduction, eye closure).

Dystonic movements are occasionally rhythmic and most often arrhythmic. When rhythmic, they are difficult to differentiate from non-dystonic essential tremor [10]. Aside from their directional character, clinical features indicating rhythmic dystonia, rather than essential tremor, include: irregularity, the presence or worsening of tremor when the affected body part is placed in a position opposite to the direction of pull, and activation of muscles not required for maintenance of the movement (overflow, as described below). By contrast, dystonic movements are easily distinguished from chorea: in dystonia there is no

flowing of movement along the affected body parts, and muscle tone is not reduced. Dystonic movements may have different speeds. When fast, they may resemble myoclonus and generate what has been termed "myoclonic dystonia" [11, 13]; when slow and distal they match the description of athetosis [14]. Unlike tics that usually change their pattern over time, dystonic movements are predicable and consistent during an observational period. Furthermore, there is no strong urge to execute the involuntary movement and no relief after execution; these are aspects of tics that make them assume a "semi-voluntary" or at least intentional nature not usually observed in dystonia. In people with excessive blinking it may be difficult to recognize the features of dystonic movements from those of motor tics. The expression "dystonic tics" has been used to indicate motor eye tics, which look like mild dystonia [15].

Additional Clinical Features

Dystonia is not a static phenomenon. Changes in the pattern of muscle activation occur during the course of the disease and also following specific manoeuvres that have diagnostic value [16]. The possibility to aggravate or relieve dystonia by using specific physical signs has diagnostic value in many cases.

Eliciting or Worsening Dystonia

Overflow is observed when dystonia extends to a contiguous body region where it is not observed as an independent phenomenon. An example is overflow to the upper limbs in patients with cervical dystonia. Mirroring occurs when, during a voluntary task involving a limb, similar albeit involuntary movements (often with dystonic features) arise in the contralateral limb. Mirroring is not a specific feature of dystonia, although it may reveal a latent dystonia, particularly in subjects belonging to dystonia families. When occurring in dystonic patients, mirroring can be considered as a minimal expression of focal dystonia that is observed in otherwise unaffected body regions.

The term "action dystonia" indicates that dystonia is activated by a voluntary task. Activation by voluntary movements allows to detect dystonia when it is not observed at rest or to increase its intensity when it is too mild to be unequivocally recognised. The activating voluntary movement may vary from non-specific to highly task-specific. Occupational dystonia occurs when a specific occupation (i.e., a motor task) is performed. Task specificity is a feature of mild forms of dystonia, which may be lost with progression.

Primary writing tremor, was first described in a patient who complained of jerking of the right forearm on writing [17]. Despite its name, this is considered a task-specific dystonia where the movement resembles tremor due to its rhythmicity. Similarly to dystonic movements, dystonic postures may also be activated by specific voluntary motor tasks.

Transiently Improving Dystonia

Dystonic movements and postures may be alleviated by some specific voluntary movements, also called gestes antagonistes, or by sensory tricks [18, 19]. Their observation strongly supports the diagnosis of dystonia [20]. They are thought to inhibit, at the central level, the cortical overflow associated with dystonia [21] and their finding is a clinical sign for the diagnosis of dystonia.

The two terms hint at different pathophysiology: performing a highly specific voluntary movement may interfere with the outflow of motor programmes from the basal ganglia thus inhibiting dystonia (gestes antagonistes); on the other hand, sensory afferents may inhibit the clinical emergence of dystonia (sensory tricks) [22]. Thus, gestes antagonistes and sensory tricks do not merely counteract the involuntary movement.

Patients often automatically select gestes antagonistes when dystonic movements are at their peak; this can be regarded as an exception to the general rule that voluntary movements, particularly purposeful skilled actions, aggravate dystonia, either mobile or fixed. For that reason, the amelioration of dystonia with activity has also been termed "paradoxical dystonia" [23]. The usage of this confounding expression is, however, discouraged. The gestes antagonistes normally involve a body part different from (and often contiguous with) the one affected by dystonia that is alleviated.

Recommendations and Good Practice Points

Clinical Diagnosis

- Diagnosis and classification of dystonia are highly relevant for providing appropriate management, prognostic information, genetic counselling and treatment (good practice point).
- Based on the lack of specific diagnostic tests, expert observation is recommended. Referral to a movement disorders expert increases the diagnostic accuracy [24] (good practice point).
- Neurological examination alone allows the clinical identification of primary dystonia and dystonia plus, but not the distinction among different etiological forms of heredo-degenerative and secondary dystonias (good practice point).

Laboratory Tests

- Diagnostic DYT1 testing in conjunction with genetic counseling is recommended for patients with primary dystonia with onset before age 30 [25] (level B).
- Diagnostic DYT1 testing in patients with onset after age 30 may also be warranted in those having an affected relative with early onset [25, 26] (level B).
- Diagnostic DYT1 testing is not recommended in patients with onset of symptoms after age 30 who either have focal cranial-cervical dystonia or have no affected relative with early onset dystonia [25, 26] (level B).
- Diagnostic DYT1 testing is not recommended in asymptomatic individuals, including those under the age of 18, who are relatives of familial dystonia patients. Positive genetic testing for dystonia (e. g. DYT1) is not sufficient to make a diagnosis of dystonia unless clinical features show dystonia [25, 27] (level B).
- A diagnostic levodopa trial is warranted in every patient with early onset dystonia without an alternative diagnosis [28] (good practice point).
- Individuals with myoclonus affecting the arms or neck, particularly if positive for autosomal dominant inheritance, should be tested for the DYT11 gene [29] (good practice point).
- Neurophysiological tests are not routinely recommended for the diagnosis or classification of dystonia; however, the observation of abnormalities typical of dystonia is an additional diagnostic tool in cases where the clinical features are considered insufficient to the diagnosis [30, 31] (good practice point).
- Brain imaging is not routinely required when there is a confident diagnosis of primary dystonia in adult patients, because a normal study is expected in primary dystonia [32] (good practice point).
- Brain imaging is necessary for screening of secondary forms of dystonia, particularly in the pediatric population due to the more widespread spectrum of dystonia at this age [33] (good practice point).

- MRI is preferable to CT, except when brain calcifications are suspected (good practice point).
- There is no evidence that more sophisticated imaging techniques (e. g., voxel-based morphometry, DWI, fMRI) are currently of any value in either the diagnosis or the classification of dystonia (good practice point).

Treatment

Botulinum Toxins

- Botulinum toxin type A (BoNT-A; or type B if there is resistance to type A) can be regarded as first line treatment for primary cranial (excluding oromandibular) or cervical dystonia [34, 35] (level A).
- Due to the large number of patients who require BoNT injections, the burden of performing treatment could be shared with properly trained nurse specialists, except in complex dystonia or where EMG guidance is required [36] (level B).
- BoNT-A may be considered in patients with writing dystonia [37] (level C).

Anticholinergic Drugs

The absolute and comparative efficacy and tolerability of anticholinergic agents in dystonia is poorly documented in children and there is no proof of efficacy in adults; therefore, no recommendations can be made to guide prescribing (good practice point).

Antiepileptic Drugs

There is lack of evidence to give recommendations for this type of treatment (good practice point).

Anti-Dopaminergic Drugs

There is lack of evidence to give recommendations for this type of treatment (good practice point).

Dopaminergic Drugs

Following a positive diagnostic trial with levodopa, chronic treatment with levodopa should be initiated and adjusted according to the clinical response [38] (good practice point).

Neurosurgical Procedures

- Pallidal DBS is considered a good option, particularly for generalized or cervical dystonia, after

medication or BoNT have failed to provide adequate improvement. While it can be considered second-line treatment in patients with generalized dystonia, this is not the case in cervical dystonia since there are other surgical options available (see below). This procedure requires a specialized expertise, and is not without side effects [39, 40] (good practice point).
- Selective peripheral denervation is a safe procedure with infrequent and minimal side effects that is indicated exclusively in cervical dystonia. This procedure requires a specialized expertise [41] (level C).
- There is insufficient evidence to use intrathecal baclofen in primary dystonia; the procedure can be indicated in patients where secondary dystonia is combined with spasticity [42] (good practice point).
- Radiofrequency lesions are currently discouraged for bilateral surgery [43] (good practice point).

Conclusions and Implications for Clinical Practice Today

Diagnosis and classification of dystonia are highly relevant for providing appropriate management and prognostic information, and genetic counseling. Expert observation is suggested. DYT1 gene testing is recommended for patients with primary dystonia with onset before age 30 and in those with an affected relative with early onset. Positive genetic testing for dystonia (e. g. DYT1) is not sufficient to make a diagnosis of dystonia.

BoNT-A (or type B if there is resistance to type A) can be regarded as first-line treatment for primary cranial (excluding oromandibular) or cervical dystonia and can be effective in writing dystonia. Pallidal deep brain stimulation (DBS) is considered a good option, particularly for generalized or cervical dystonia, after medication or BoNT have failed to provide adequate improvement. Selective peripheral denervation is a safe procedure that is indicated exclusively in cervical dystonia.

Future Directions

The Task Force will meet regularly to review new evidence on diagnosis and treatment and will update its recommendations according to emerging new scientific evidence.

References

1. Fahn S, Marsden CD, Calne DB. Classification and investigation of dystonia. In: Marsden CD, Fahn S, eds. Movement disorders 2. London: Butterworths, 1987: 332–358
2. Fahn S, Bressman S, Marsden CD. Classification of dystonia. Adv Neurol 1998; 78: 1–10
3. Defazio G, Abbruzzese G, Livrea P, Berardelli A. Epidemiology of primary dystonia. Lancet Neurol 2004; 3: 673–678
4. Brainin M, Barnes M, Baron JC et al. Guidance for the preparation of neurological management guidelines by EFNS scientific task forces – revised recommendations 2004. Eur J Neurol 2004; 11: 577–581
5. Deuschl G, Bain P, Brin M. Consensus statement of the Movement Disorder Society on Tremor. Ad Hoc Scientific Committee. Mov Disord 1998; 13 (Suppl 3): 2–23
6. Hallett M, Daroff RB. Blepharospasm: report of a workshop. Neurology 1996; 46: 1213–1218
7. Sanger TD, Delgado MR, Gaebler-Spira D, Hallett M, Mink JW. Classification and definition of disorders causing hypertonia in childhood. Pediatrics 2003; 111: e89–e97
8. Albanese A. The clinical expression of primary dystonia. J Neurol 2003; 250: 1145–1151
9. Chan J, Brin MF, Fahn S. Idiopathic cervical dystonia: clinical characteristics. Mov Disord 1991; 6: 119–126
10. Munchau A, Schrag A, Chuang C et al. Arm tremor in cervical dystonia differs from essential tremor and can be classified by onset age and spread of symptoms. Brain 2001; 124: 1765–1776
11. Davidenkow S. Auf hereditär-abiotrophischer Grundlage akut auftretende, regressierende und episodische Erkrankungen des Nervensystems und Bemerkungen über die familiäre subakute, myoklonische Dystonie. Z Ges Neurol Psychiat 1926; 104: 596–622
12. Obeso JA, Rothwell JC, Lang AE, Marsden CD. Myoclonic dystonia. Neurology 1983; 33: 825–830
13. Benedek L, Rakonitz E. Heredopathic combination of a congenital deformity of the nose and of myoclonic torsion dystonia. J Nerv Ment Dis 1940; 91: 608–616
14. Hammond WA. A treatise of diseases of the nervous system. New York: Appleton Century Crofts, 1871
15. Jankovic J. Botulinum toxin in the treatment of dystonic tics. Mov Disord 1994; 9: 347–349
16. Gelb DJ, Yoshimura DM, Olney RK, Lowenstein DH, Aminoff MJ. Change in pattern of muscle activity following botulinum toxin injections for torticollis. Ann Neurol 1991; 29: 370–376
17. Rothwell JC, Traub MM, Marsden CD. Primary writing tremor. J Neurol Neurosurg Psychiatry 1979; 42: 1106–1114
18. Greene PE, Bressman S. Exteroceptive and interoceptive stimuli in dystonia. Mov Disord 1998; 13: 549–551
19. Gomez-Wong E, Marti MJ, Cossu G, Fabregat N, Tolosa ES, Valls-Sole J. The 'geste antagonistique' induces transient modulation of the blink reflex in human patients with blepharospasm. Neurosci Lett 1998; 251: 125–128
20. Masuhr F, Wissel J, Muller J, Scholz U, Poewe W. Quantification of sensory trick impact on tremor amplitude and frequency in 60 patients with head tremor. Mov Disord 2000; 15: 960–964
21. Hallett M. Is dystonia a sensory disorder? Ann Neurol 1995; 38: 139–140
22. Kaji R, Rothwell JC, Katayama M, et al. Tonic vibration reflex and muscle afferent block in writer's cramp. Ann Neurol 1995; 38: 155–162
23. Fahn S. Clinical variants of idiopathic torsion dystonia. J Neurol Neurosurg Psychiatry 1989; Special Suppl: 96–100
24. Logroscino G, Livrea P, Anaclerio D et al. Agreement among neurologists on the clinical diagnosis of dystonia at different body sites. J Neurol Neurosurg Psychiatry 2003; 74: 348–350
25. Klein C, Friedman J, Bressman S et al. Genetic testing for early-onset torsion dystonia (DYT1): introduction of a simple screening method, experiences from testing of a large patient cohort, and ethical aspects. Genet Test 1999; 3: 323–328
26. Bressman SB, Sabatti C, Raymond D et al. The DYT1 phenotype and guidelines for diagnostic testing. Neurology 2000; 54: 1746–1752
27. Points to consider: ethical, legal, and psychosocial implications of genetic testing in children and adolescents. American Society of Human Genetics Board of Directors, American College of Medical Genetics Board of Directors. Am J Hum Genet 1995; 57: 1233–1241
28. Robinson R, McCarthy GT, Bandmann O, Dobbie M, Surtees R, Wood NW. GTP cyclohydrolase deficiency; intrafamilial variation in clinical phenotype, including levodopa responsiveness. J Neurol Neurosurg Psychiatry 1999; 66: 86–89
29. Valente EM, Edwards MJ, Mir P et al. The epsilon-sarcoglycan gene in myoclonic syndromes. Neurology 2005; 64: 737–739
30. Deuschl G, Heinen F, Kleedorfer B, Wagner M, Lucking CH, Poewe W. Clinical and polymyographic investigation of spasmodic torticollis. J Neurol 1992; 239: 9–15
31. Hughes M, McLellan DL. Increased co-activation of the upper limb muscles in writer's cramp. J Neurol Neurosurg Psychiatry 1985; 48: 782–787
32. Rutledge JN, Hilal SK, Silver AJ, Defendini R, Fahn S. Magnetic resonance imaging of dystonic states. Adv Neurol 1988; 50: 265–275
33. Meunier S, Lehericy S, Garnero L, Vidailhet M. Dystonia: lessons from brain mapping. Neuroscientist 2003; 9: 76–81
34. American Academy of Ophthalmology. Botulinum toxin therapy of eye muscle disorders. Safety and effectiveness. Ophthalmology 1989; 96 (Pt 2): 37–41
35. Costa J, Espirito-Santo C, Borges A, et al. Botulinum toxin type A therapy for blepharospasm. Cochrane Database Syst Rev 2005; CD004900
36. Whitaker J, Butler A, Semlyen JK, Barnes MP. Botulinum toxin for people with dystonia treated by an outreach nurse practitioner: a comparative study between a home and a clinic treatment service. Arch Phys Med Rehabil 2001; 82: 480–484
37. Balash Y, Giladi N. Efficacy of pharmacological treatment of dystonia: evidence-based review including meta-analysis of the effect of botulinum toxin and other cure options. Eur J Neurol 2004; 11: 361–370
38. Hwang WJ, Calne DB, Tsui JK, Fuente-Fernandez R. The long-term response to levodopa in dopa-responsive dystonia. Parkinsonism Relat Disord 2001; 8: 1–5
39. Vidailhet M, Vercueil L, Houeto JL, et al. Bilateral deep-brain stimulation of the globus pallidus in primary generalized dystonia. N Engl J Med 2005; 352: 459–467

40. Eltahawy HA, Saint-Cyr J, Poon YY, Moro E, Lang AE, Lozano AM. Pallidal deep brain stimulation in cervical dystonia: clinical outcome in four cases. Can J Neurol Sci 2004; 31: 328–332

41. The National Institute for Clinical Excellence. Selective peripheral denervation of cervical dystonia. 2004

42. Albright AL, Barry MJ, Shafton DH, Ferson SS. Intrathecal baclofen for generalized dystonia. Dev Med Child Neurol 2001; 43: 652–657

43. Loher TJ, Pohle T, Krauss JK. Functional stereotactic surgery for treatment of cervical dystonia: review of the experience from the lesional era. Stereotact Funct Neurosurg 2004; 82: 1–13

Complex Regional Pain Syndrome – What's in a Name?

B. van Hilten

Introduction

In 1994 the term Complex Regional Pain Syndrome (CRPS) was introduced along with its criteria, which focused on sensory and autonomic features of this disorder [1]. These criteria were to replace the term "reflex sympathetic dystrophy" (RSD) with CRPS type I and causalgia with CRPS type II, respectively [1]. The difference between the two types of CRPS is based on the absence (CRPS type I) or presence (CRPS type II) of an overt nerve lesion. CRPS frequently follows tissue injury, which can be minimal or severe (sprain/strain, fracture, contusion/crush injury) [2]. But in 5–16% of the patients, no inciting event can be identified. As with prior RSD criteria sets, the CRPS criteria of the IASP focus on the different aspects of sensory and autonomic features [3, 4]. However, there is a growing recognition that the clinical spectrum of CRPS is broader including also movement disorders. Additionally, the CRPS criteria set has a low specificity [5]. Recent reviews on the utilisation of diagnostic criteria in studies/trials on CRPS highlight a lack of consensus on the content and application of criteria sets [4–6]. Together, these short-comings have led to new criteria that were published in 2005 (see below) [7].

Modified IASP research diagnostic criteria for CRPS-1 – Budapest criteria [2] (submitted to Committee for Classification of Chronic Pain of the IASP for the 3rd taxonomy, not yet accepted):

1. Continuing pain, which is disproportionate to any inciting event.
2. Must report at least one symptom in each of the four following categories:
   - Sensory: reports of hyperesthesia and/or allodynia.
   - Vasomotor: reports of temperature asymmetry and/or skin color changes and/or skin color asymmetry.
   - Sudomotor/edema: reports of edema and/or sweating changes and/or sweating asymmetry.
   - Motor/trophic: reports of decreased range of motion and/or motor dysfunction (weakness, tremor, dystonia) and/or trophic changes (hair, nails, skin).
3. Must display at least one sign* in two or more of the following categories:
   - Sensory: evidence of hyperalgesia (to pinprick) and/or allodynia (to light touch and/or deep somatic pressure and/or joint movement)
   - Vasomotor: e.vidence of temperature asymmetry and/or skin color changes and/or asymmetry.
   - Sudomotor/edema: evidence of edema and/or sweating changes and/or sweating asymmetry.

* A sign is counted only if observed at the time of diagnosis.

- Motor/trophic: evidence of decreased range of motion and/or motor dysfunction (weakness, tremor, dystonia) and/or trophic changes (hair, nails, skin).
4. There is no other diagnosis that better explains the signs and symptoms.

The Clinical Spectrum of CRPS: How Complex?

Defining the typical spectrum of CRPS is a challenging task (because symptoms and signs can be difficult to identify), may occur in different combinations, and vary over time. Key features in the acute phase of CRPS are characterized by various combinations of sensory and autonomic symptoms and signs [3]. Although, the CRPS criteria require the presence of pain, this has been a controversial issue. In the series of Veldman, 4–7% of the cases had prominent autonomic symptoms and signs but no pain [3]. Additionally, while there has always been a focus on the sensory and autonomic features, there is increasing evidence that movement disorders are part of the spectrum of CRPS. Movement disorders may precede the occurrence of non-motor features of CRPS [3, 8, 9]. Some studies have even highlighted the sole occurrence of dystonia or tremor following trauma [10–13]. These movement disorders are concordant with those encountered in CRPS patients that suffer from sensory and autonomic features as well. The movement disorders occurring in CRPS patients may include weakness, dystonia, tremor and myoclonus, but frequently different combinations of these movement disorders may coincide within one patient [3, 8, 14, 15]. There is no reliable information on the incidence and prevalence of the different movement disorders in CRPS because epidemiological studies on CRPS suffer from methodological shortcomings, including selection bias (data are obtained from university-based tertiary chronic pain clinics or trauma units), design, and anecdotal reports. Nevertheless, the increasing awareness that CRPS patients may suffer from movement disorders has resulted in adding this clinical category to the new criteria set (see above) [7]. Many studies have documented the presence of weakness or a limited range of motion. However, both are not necessarily motor features as they may result from pain, edema or arthrogenic changes. A frequent finding in CRPS patients with weakness and/or dystonia is the so-called loss of voluntary control; patients will describe this phenomena as "My mind tells my hand to move, but it won't work" [8, 16, 17]. The loss of voluntarily control has also been reported in primary dystonia [18]. In our experience, bradykinesia is a typical abnormal movement characteristic in CRPS, even in patients that solely suffer from pain. Dystonia, a prominent motor feature of CRPS, is characterized by involuntary abnormal, predominant flexor postures (fixed dystonia) of the fingers, wrist and feet [8, 15]. In more severely affected patients the dystonia may progress to more proximal sites with again predominant flexor involvement [15]. Clinically, fixed dystonia may show a variable degree of flexion of the digits. In less affected patients, the hands may appear seemingly normal at inspection. In these patients, dystonia may only appear following the performance of repetitive tasks. In many patients there is a relative sparing of the first two digits that has been explained by a larger proportion of direct cortico-motoneuronal connections relative to interneuronal-motoneuronal connections of flexors of digits I and II [15]. Hence, the preferential involvement of flexors III–V has been interpreted as evidence pointing towards a role of abnormal function of interneuronal circuits [15]. Passive stretching of the affected digits results in a contraction of the stretched muscle suggesting a stretch reflex hyperexcitability [15, 17]. Dystonia may worsen by activity of the involved extremity, under circumstances of cold temperatures and humidity and in the more severely affected patients by tactile and auditory stimuli [15]. Myoclonus and tremor (3–7 Hz) are frequently reported by CRPS patients with dystonia, but in rare cases this may occur as the sole or predominant movement disorder [8, 15, 18, 20].

CRPS: How Regional?

CRPS is commonly known as a disorder affecting one extremity. However, several studies have highlighted that in 4–7% of the cases, the disease may spread to other extremities [15, 20, 21]. The spread of CRPS may result in rather unusual patterns characterized by multifocal or generalized distribution [15, 20, 21]. These more severely affected patients tend to be younger than those patients where CRPS remains restricted to one extremity [20, 21].

In more severely affected CRPS patients where the disease has spread to other extremities, it is not unusual to encounter bladder (urgency, retention) and bowel (obstipation, diarrhoea, or a combination of both) disorders as manifestations of CRPS [22, 23].

CRPS: Multiple Underlying Pathophysiological Mechanisms?

Although several hypotheses have been suggested, including sympathetic hyperactivity, changes in adrenergic sensitivity and psychological predisposition, the pathophysiological basis of CRPS is still unclear. Similarities between the classical symptoms of inflammation and the clinical features of CRPS have led several investigators to suggest an inflammatory origin of the disease [24, 25]. Indeed, the evidence pointing towards a possible involvement of the peripheral nervous system in the generation of inflammatory response in CRPS is compelling. CRPS has not been reported in patients with complete nerve lesions, suggesting that at least some continuity of a nerve is a pre-requisite to develop this disorder [26]. Both sensory and autonomic symptoms of CRPS occur in a similar glove- or stocking-like distribution pattern pointing towards a common underlying mechanism of both features [15]. Increasingly, research is documenting a perturbed function of C- and Aδ-fibres of sensory nerves as a potentially important candidate mechanism in the acute phase of CRPS [27–30]. Besides warning us of imminent or actual tissue damage of the skin, C and Aδ-fibres of sensory nerves respond to this damage as a first line of defence through the release of the neuropeptides substance P and Calcitonin gene-related peptide (CGRP) from the afferent nerve endings, a process known as neurogenic inflammation [31, 32]. This in turn results in local vasodilatation and increased capillary permeability causing edema and an increase of skin blood flow. Although peripheral nerve involvement is likely to play an important role in the acute phase of CRPS, numerous abnormalities on the neuroimmune level may play a role as well.

In contrast to the sensory and autonomic features, movement disorders in CRPS tend to become more prevalent as the disease duration lengthens [3]. Consequently, this suggests that a different mechanism may underlie the occurrence of movement disorders in CRPS. Most likely, this mechanism reflects the development of altered sensory motor integration on the spinal cord level as has been noted for peripheral nerve lesions [33–35]. The ability to experience pain serves a purpose as noxious stimuli elicit protective withdrawal reflexes, which generally involve flexor muscles to minimize or avoid potential tissue damage. The conspicuous involvement of flexor muscles in CRPS patients who have dystonia therefore hints towards the involvement of spinal motor programs that are involved in protective responses against pain [36].

Neurophysiological studies have revealed impairment of interneuronal circuits that mediate presynaptic inhibition of motoreurons of distal musculature and postsynaptic inhibition of motoneurons of proximal musculature [17, 37]. Successful pharmacological treatment of dystonia of CRPS by means of intrathecal administration of baclofen, a GABA B agonist, has highlighted the involvement of spinal GABAergic inhibitory interneurons [38]. These interneurons inhibit the amount of excitatory synaptic transmitter released by the sensory input on motoneurons in the spinal cord by means of presynaptic inhibition [39]. Spinal GABAergic interneurons receive both inputs from sensory nerves and descending fibres from the brainstem and motor cortex (supraspinal), and therefore have a strategic position in the regulation of muscle tone [39, 40]. Through impairment of these interneurons, motoneurons are exposed to an uninhibited sensory and supraspinal input explaining the worsening of dystonia by tactile stimuli, low temperatures, activity of the involved extremity, and emotional stress. Taken together, the above findings on fixed dystonia of CRPS are in line with the general pathophysiological concept of abnormal central sensorimotor processing in primary and secondary dystonia [38, 39].

Central sensitization is an important mechanism in pain and reflects the increased sensitivity of spinal neurons, despite unchanged afferent input. As a result, pain becomes chronic, and non-noxious stimuli become painful [44]. On a molecular level, central sensitization is associated with changes in the release of neuropeptides, neurotransmitters, prostaglandine E2, and the expression of N-methyl aspartate (NMDA) receptors [44, 45]. In view of the mechanisms by which they evolve and the time frame in which they appear, the movement disorders of CRPS most likely evolve within the context of central sensitization [36]. Although, the mechanism underlying the impairment of GABAergic inhibitory interneurons in CRPS is unknown; there are indications that substance P may mediate these changes [46, 47].

CRPS: A Multifactorial Disease?

CRPS has many characteristics that are typical of multifactorial disease. On the one hand, in CRPS a wide range of precipitating trauma has been identified [3]. In response to trauma, the body responds with a series of specific reactions aiming to repair the damage, promote wound healing and recruit host

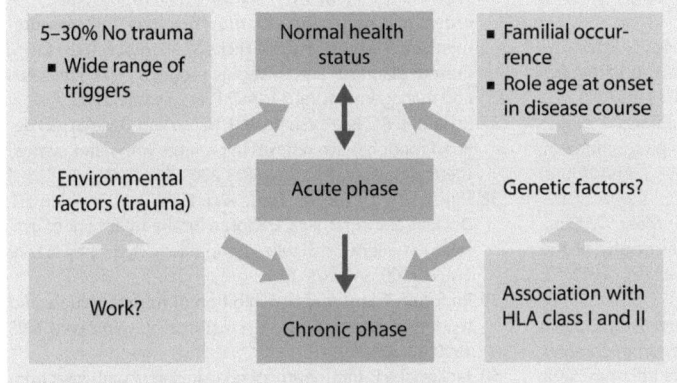

◘ Fig. 2.1. Mechanisms to trauma

defence mechanisms that involve bi-directionally acting components of the nervous system and the immune system [48]. In CRPS, this biological defence mechanism to noxious or non-noxious stimuli apparently has the capacity of becoming detrimental when it cannot be controlled appropriately. On the other hand, as indicated by a younger age at onset in cases with a progressive disease course and the association with HLA factors, there is evidence suggesting a role for genetic factors conferring susceptibility to develop or sustain CRPS [23, 49]. Taken together, CRPS likely stands as an intriguing human model of aberrant response mechanisms to trauma (◘ Fig. 2.1).

References

1. Merskey H, Bogduk N. Complex regional pain syndrome, type I (Reflex sympathetic dystrophy). In: Classification of chronic pain: descriptions of chronic pain syndromes and definitions of pain terms, 2nd edn. IASP Press, Seattle, WA, 1994, pp 41–42

2. Allen G, Galer BS, Schwartz L. Epidemiology of complex regional pain syndrome: a chart retrospective review of 134 patients. Pain 1999; 539–544

3. Veldman PH, Reynen HM, Arntz IE, Goris RJ. Signs and symptoms of reflex sympathetic dystrophy: prospective study of 829 patients. Lancet 1993; 342: 1012–1016

4. Harden RN, Bruehl S, Galer BS et al. Complex regional pain syndrome: are the IASP diagnostic criteria valid and sufficiently comprehensive? Pain 1999; 83: 211–219

5. Bruehl S, Harden RN, Galer BS. External validation of the IASP diagnostic criteria for Complex Regional pain syndrome and proposed research diagnostic criteria. Pain 1999; 81:147–154

6. Beek van de WJT, Schwartzman, Nes van SI, Delhaas EM, Hilten van JJ. Diagnostic criteria utilized in studies of reflex sympathetic dystrophy. Neurology 2002; 58: 522–526

7. Harden NR, Bruehl S. Diagnostic Criteria: The statistical derivation of the four criterion factors. In: Wilson P, Stanton-Hicks M, Harden RN (eds) CRPS: Current diagnosis and therapy, progress in pain research and management, Vol 32. IASP Press, Seattle, 2005

8. Schwartzman RJ, Kerrigan J. The movement disorder of reflex sympathetic dystrophy. Neurology 1990;40:57–61

9. Steinbrocker O. The shoulder-hand syndrome in reflex dystrophy of the upper extremity. Ann Intern Med 1948; 29: 22–52

10. Jankovic J, Van der Linden C. Dystonia and tremor induced by peripheral trauma: predisposing factors. J Neurol Neurosurg Psychiatry 1988; 51: 1512–1519

11. Schott GD. Induction of involuntary movements by peripheral trauma: an analogy with causalgia. Lancet 1986; 8509: 712–716

12. Fletcher NA, Harding AE, Marsden CD. The relationship between trauma and idiopathic torsion dystonia. J Neurol Neurosurg Psychiatry 1991; 54: 713–717

13. Cardoso F, Jankovic J. Peripherally induced tremor and parkinsonism. Arch Neurol 1995; 52: 263–270

14. Birklein F, Riedl B, Sieweke N, Weber M, Neurndorfer B. Neurologic findings in complex regional pain syndromes: analysis of 145 cases. Acta Neurol Scand 2000a; 101: 262–269

15. Hilten van JJ, van de Beek WJ, Vein AA, van Dijk JG, Middelkoop HA. Clinical aspects of multifocal or generalized tonic dystonia in reflex sympathetic dystrophy. Neurology 2001; 56: 1762–1765

16. Galer BS, Jensen M. Neglect-like symptoms in complex regional pain syndrome: results of a self-administered survey. J Pain Symptom Manage 1999; 18: 213–217

17. Beek van de WJT, Vein A, Hilgevoord AAJ, Dijk van JG, Hilten van JJ. Neurophysiological aspects of patients with generalized or multifocal dystonia in reflex sympathetic dystrophy. J Clin Neurophysiology 2002; 19: 77–83

18. Berardelli A, Rothwell JC, Hallet M et al. The pathophysiology of primary dystonia. Brain; 1998; 121: 1195–1212

19. Deuschl G, Blumberg H, Lucking CH. Tremor in reflex sympathetic dystrophy. Arch Neurol 1991; 48: 1247–1252

20. Van der Laan, Veldman PH, Goris RJ. Severe complications of reflex sympathetic dystrophy: infection, ulcers, edema, dystonia and myoclonus. Arch Phys Med Rehabil 1998; 79:424–429

21. Bhatia KP, Bhatt MH, Marsden CD. The Causalgia-dystonia syndrome. Brain 1993; 116: 843–851

22. Chancellor MB, Shenot PJ, Rivas DA, Mandel, S, Schwartzman RJ, Urological symptomatology in patients with reflex sympathetic dystrophy. J Urol 1996; 155: 634–637

23. Hilten van JJ, van de Beek WJT, Roep B. Multifocal or generalized tonic dsytonia of Complex Regional pain Syndroem: A distinct entity associated with HLA-DR-13. Ann Neurol 2000; 48: 113–116

24. Oyen WJ, Arntz IE, Claessens RM, Van der Meer JW, Corstens FH, Goris RJ. Reflex sympathetic dystrophy of the hand: an excessive inflammatory response? Pain 1993; 55: 151–157

25. Huygen FJ, De Bruijn AG, De Bruin MT, Groeneweg JG, Klein J, Zijlstra FJ. Evidence for local inflammation in complex regional pain syndrome type 1. Mediators Inflamm 2002; 11: 47–51

26. Lewis T. The nocifensor system of nerves and its reactions. BMJ 1937; 1: 431–435

27. Blair SJ, Chinthagada M, Hoppenstehdt D, Kijowski R, Fareed J. Role of neuropeptides in pathogenesis of reflex sympathetic dystrophy. Acta Orthop Belg 1998; 64: 448–451

28. Weber M, Birklein F, Neundorfer B, Schmeltz M. Facilitated neurogenic inflammation in complex regional pain syndrome. Pain 2001; 91: 251–257

29. Birklein F, Schmelz M, Schifter S, Weber M. The important role of neuropeptides in complex regional pain syndrome. Neurology 2001; 26: 2179–2184

30. Leis S, Weber M, Isselmann A, Schmelz M, Birklein F. Substance-P-induced protein extravasation is bilaterally increased in complex regional pain syndrome. Exp Neurol. 2003; 183:197–204

31. Holzer P, Maggi CA. Dissociation of dorsal root ganglion neurons into afferent and efferent-like functions. Neuroscience 1998; 86: 389–398

32. Brain SD, Moore PK (eds). Pain and neurogenic inflammation. Birkhäuser, Basel, 1999

33. Girlanda P, Quartarone A, Battaglia F, Pcciolo G, Sinicropi S, Messina C. Changes in spinal cord excitability in patients affected by ulnar neuropathy. Neurology 2000; 55: 975–978

34. Merzenich MM, Kaas JH, Wall JT, Sur M, Nelson RJ, Fellemann DJ. Progression of changes following median nerve section in the cortical representation of the hand in areas 3b and 1 in adult owl and squirrel monkey. Neuroscience 1983; 10: 639–665

35. Cohen LG, Bandinelli S, Findley TW, Hallet M. Motor reorganization after upper limb amputation in man. A study with focal magnetic stimulation. Brain 1991; 114:615–627

36. Hilten van JJ, Blumberg H, Schwartzman RJ. Movement disorders and dystrophy – pathophysiology and measurement. In: Wilson P, Stanton-Hicks M, Harden N (eds) CRPS: Current diagnosis and therapies, progress in pain research and management, Vol 32. IASP Press, Seattle, 2005

37. Schouten AC, Beek van de WJT, Hilten van JJ, Helm van der FCT. Proprioceptive reflexes in patients with reflex sympathetic dystrophy. Exp Brain Res 2003; 151: 1–8

38. Hilten van JJ, van de Beek WJT, Hoff JI, Voormolen JH, Delhaas EM. Intrathecal baclofen for the treatment of dystonia in patients with reflex sympathetic dystrophy. N Engl J Med 2000; 343: 625–630

39. Rudomin P. Presynaptic inhibition of muscle spindle and tendon organ afferents in the mammalian spinal cord. TINS 1990; 13: 499–505

40. Jankowska E. Interneuronal relay in spinal pathways from proprioceptors. Prog Neurobiol 1992; 38:335–378

41. Hallett M. Is dystonia a sensory disorder? Ann Neurol. 1995; b38: 139–140

42. Tamburin S, Zanette G. Abnormalities of sensory processing and sensorimotor interactions in secondary dystonia: a neurophysiological study in two patients. Mov Disord 2005; 20: 354–360

43. Woolf C, Wiesenfeld-Hallin Z. Substance P and calcitonin gene-related peptide synergistically modulate the gain of the nociceptive flexor withdrawal reflex in the rat. Neurosci Lett 1986; 66: 226–30

44. Woolf CJ, Mannion RJ. Neuropathic pain: aetiology, symptoms, mechanisms, and management. Lancet 1999; 353: 1959–1964

45. Samad TA, Moore KA, Sapirstein A, Billet S, Allchorne A, Poole S, Bonventre JV, Woolf CJ. Interleukin-1b-mediated induction of Cox-2 in the CNS contributes to inflammatory pain hypersensitivity. Nature 2001; 410: 471–475

46. Woolf C, Wiesenfeld-Hallin Z. Substance P and calcitonin gene-related peptide synergistically modulate the gain of the nociceptive flexor withdrawal reflex in the rat. Neurosci Lett 1986; 15; 66: 226–230

47. Parsons AM, Honda CN, Jiay P et al. Spinal NK1 receptors contribute to the increased excitability of the nociceptive flexor reflex during persistent peripheral inflammation. Brain Res 1996; 739: 263–275

48. Black PH. Stress and the inflammatory response: a review of neurogenic inflammation. Brain Behav Immun 2002; 16: 622–653

49. van de Beek WJ, Roep BO, van der Slik AR, Giphart MJ, van Hilten BJ. Susceptibility loci for complex regional pain syndrome. Pain 2003; 103: 93–97

Neuropathic Pain: Assessment and Medical Treatment

G. Cruccu

Definition and Classification

Pain specialists feel that the major problems with neuropathic pain regard its treatment. We also believe that definition/classification and diagnostic tools are currently insufficient.

It is generally understood, however, that the lesion must involve the somatosensory pathways with damage to small fibers in peripheral nerves or to the spino-thalamo-cortical system in the CNS. Previous classifications of neuropathic pain have been based on underlying disease (e. g. diabetic neuropathy, multiple sclerosis etc.) or site of lesion (e. g. peripheral nerve, spinal cord etc.). Traditionally, neurologists have considered neuropathic pains to be present only when there are definite signs of a nervous lesion. The issue about definition became even more demanding following the suggestion of a mechanism-based classification (Woolf and Max 2001). Some characteristics of neuropathic pain such as sensitized nociceptors, allodynia, abnormal temporal summation, or extra-territorial spread of pain, are also shared by less clear chronic pain conditions (Hansson et al. 2001; Jensen et al. 2001). The inclusion of the word "dysfunction" in the definition of neuropathic pain implies that other conditions such as complex regional pain syndromes or even musculo-skeletal disorders associated with signs of hypersensitivity may be considered neuropathic pains. ◘ Table 3.1 provides an example of the difficulties currently met in classification.

At present, a task force of the IASP taxonomic committee is preparing a new definition and a diagnostic grading as "definite, probable, possible" neuropathic pain. In the meanwhile, we suggest that the narrow definition and classification is retained, because of the risk of overestimating neuropathic pain and because it is easy to understand.

Some conditions are listed in the putative and/or mixed groups to emphasize the difficulties currently met in classification, until new pathophysiological evidence and more precise definitions are achieved.

Bedside Examination

The examination of a pain patient aims at clarifying underlying disease and understanding whether the pain is nociceptive, neuropathic, psychogenic, or a combination of these. In ca ase of neuropathic pain, abnormal sensory findings should be neuroanatomically logical, compatible with a definite lesion site. Location, quality, and intensity of pain should be assessed. A clear understanding of the possible types of negative (e. g. sensory loss) and positive (e. g. paresthesia) symptoms and signs is necessary. Neuropathic pain can be spontaneous (stimulus-independent or spontaneous pain) or elicited by a stimulus (stimulus-dependent or stimulus-evoked pain). Spontaneous pain is often described as a constant burning sen-

■ Table 3.1. Informal classification (examples) of neuropathic/putative/mixed pain conditions

Examples of definite neuropathic	Examples of putative neuropathic	Examples of mixed neuropathic/nociceptive
Diabetic neuropathy	CRPS I	Lombosciatalgia/cervicobrachialgia
Small-fibre polyneuropathy	CRPS II	FBSS
HIV neuropathy	Fibromyalgia	Postmastectomy pain
Traumatic/entrapment neuropathies	Atypical facial pain	Scar pain
Phantom limb	Coccydynia/Vulvodynia	Coccydynia/Vulvodynia
Postherpetic neuralgia	Burning mouth	Burning mouth
Radiculopathy	Whiplash	Whiplash
Trigeminal neuralgia		
Spinal cord injury		
Multiple sclerosis		
Post-stroke pain		

■ Table 3.2. Summary of choice methods of assessing nerve function per sensation

Fibres	Sensation	Testing		
		Clinical	QST	Laboratory
Aβ	Touch	piece of cotton wool	von Frey filaments	Nerve conduction studies,
	Vibration	tuning fork (128 Hz)	vibrameter[a]	SEPs
Aδ	Pinprick, sharp pain	wooden cocktail stick	weighted needles	Nociceptive reflexes, LEPs
	Cold	thermorollers	thermotest[b]	no method
C	Warmth	thermorollers	thermotest[b]	LEPs
	Burning	no method	thermotest[b]	

[a]Or other device providing graded vibratory stimuli, [b]or other device providing graded thermal stimuli.
QST Quantitative Sensory Testing. SEPs Somatosensory Evoked Potentials. LEPs Laser Evoked Potentials.

sation, but may also include intermittent shooting, lancinating sensations, electric shock-like pain, and dysesthesias (i. e. abnormal, unpleasant sensations). Paresthesias are abnormal, though not unpleasant sensations. Stimulus-evoked pains are elicited by mechanical, thermal, or chemical stimuli. Hyperalgesia is an increased pain response to a stimulus that normally provokes pain, whereas allodynia is a pain sensation induced by a stimulus that normally does not provoke pain, and thus implies a change in the quality of a sensation. Mechanical allodynia, which is most easily tested, is further classified as dynamic (brush-evoked) or static (pressure-evoked). The suggested tools are summarized in ■ Table 3.2.

Quantitative Sensory Testing (QST)

Because they are also found in non-neuropathic pains, QST abnormalities cannot be taken as a conclusive demonstration of neuropathic pain (level B); furthermore QST depends on expensive equipment, it is time-consuming and thus difficult to use in clinical practice. In contrast, QST is helpful to quantify the effects of treatments on allodynia and hyperalgesia and can reveal a differential efficacy of treatments on different pain components (level B). To evaluate mechanical allodynia/hyperalgesia, we recommend the use of simple tools such as a brush and at least one high-threshold von Frey filament. The evaluation of pain in response to thermal stimuli is best performed using the thermotest, but we do not recommend the systematic measure of thermal stimuli except for pathophysiological research or treatment trials. A simple and sensitive tool to quantify pain induced by thermal stimuli in clinical practice should be developed.

Standard Electrophysiological Testing

Large-size, non-nociceptive afferents have a lower electrical threshold than small-size, nociceptive afferents. Unless special techniques are adopted (experimental blocks) or special organs are stimulated (cornea, tooth pulp, glans), electrical stimuli unavoidably also excite large, non-nociceptive afferents. The large-afferent input inhibits the nociceptive input at central synapses and hinders the nociceptive signals (IFCN Recommendations for the Practice of Clinical Neurophysiology). Hence standard neurophysiological responses to electrical stimuli, such as nerve conduction studies and somatosensory evoked potentials, are useful to demonstrate, locate, and quantify damage along the peripheral or central sensory pathways. But they do not assess function of nociceptive pathways (level B).

Laser-Evoked Potentials

For many years a number of techniques have been tried for the selective activation of pain afferents. The best method now appears to be provided by radiant-heat pulse stimuli delivered by laser stimulators, which selectively excite the free nerve endings (Aδ and C) in the superficial skin layers. That laser evoked potentials (LEPs) are nociceptive responses is now widely agreed by over 200 studies. Late LEPs reflect activity of the Aδ and ultralate LEPs of the unmyelinated nociceptive pathway (Bromm and Treede 1984, 1987, 1991; Bragard et al. 1996; Magerl et al. 1999; Cruccu et al. 2004).

Laser-evoked potentials are the easiest and most reliable neurophysiological method of assessing function of nociceptive pathways; in clinical practice their main limit is that they are currently available in too few centers. Late LEPs are diagnostically useful in peripheral and central neuropathic pains (level B). The experience as a tool for assessing treatments is so far insufficient. More studies on ultralate LEPs in patients with neuropathic pain are encouraged.

Outcome Measures

All the psychometric instruments assessing treatment in neuropathic pain have been shown sensitive in several class 2 RCTs. The simplest scales are probably the best. Whereas verbal rating scale (VRS) is found easier by many patients, visual analogue scale (VAS) is more apt to treatment trials because it permits parametric statistics. The Likert 0–10 NRS is a good compromise (level C). We recommend the use of unidimensional pain scales, particularly the numeric rating scale (NRS), the global impression of change (GIC) and the evaluation of specific pain symptoms (such as burning pain, pain paroxysms, or allodynia) since this may reveal preferential effects of treatments (level B). We do not favor the systematic use of nonspecific multidimensional scales (e. g. McGill pain questionnaire, MPQ). Although interesting, the multidimensional scales specific for neuropathic pain still lack extensive validation as tools for treatment assessment (level C).

Improvement of quality of life (QoL) has been regarded as the final aim of pain treatment. QoL is measured either by the 0–10 scale or by specific scales such as 36-item short form (SF-36; Ware et al. 1992) or the Nottingham health profile (NHP; Hunt et al. 1980; Meyer-Rosberg et al. 2001).

Medical Treatment

Selecting a first-line medication in neuropathic pain should take into account not only the relative efficacy based at best on direct drug comparisons, but also the ratio efficacy/safety. Whenever possible, the effects on pain symptoms and on comorbidities should be taken into account. However, such assessment has been performed in a limited number of studies and with only a few drugs to date, and the evaluation of pain symptoms or signs has used various methods of assessment, which were not all validated (Cruccu et al. 2004).

The effects of drugs on distinct peripheral neuropathic conditions share many similarities, with the exceptions of HIV polyneuropathy and trigeminal neuralgia. Central pain has been much less studied. For this reason, the following recommendations concern mainly the most studied neuropathic pain group, represented by peripheral neuropathic pains. Drugs with established efficacy on the basis of several class I or II trials in various etiological conditions include tricyclic antidepressants (TCAs: amitriptyline, imipramine, nortriptyline, clomipramine, desipramine, maprotiline), opioids, gabapentin, pregabalin and topical lidocaine (level A). TCAs and opioids are more efficacious on the basis of the number needed to treat (NNT). Direct comparisons based on small sample sizes have failed to demonstrate significant differences of efficacy between opioids and TCAs, and between TCAs and gabapentin, but TCAs induced more cognitive side effects than opioids in one trial (level B). Based on the ratio efficacy/safety, TCAs (particularly those with selective noradrenaline reuptake), gabapentin, pregabalin or topical lidocaine can be recommended as first line therapy. Opioids should be proposed as second-line therapy, because of a less favorable ratio efficacy/safety and precautions for use inherent to opioids in chronic non cancer pain. Antidepressants and topical lidocaine have been shown effective on various pain symptoms (level B). Effects on mechanical allodynia have been reported for lidocaine (level B) but are less documented for antidepressants. Thus the use of topical lidocaine could be preferred in patients with mechanical allodynia, especially when the area of pain is limited. Regarding comorbidities and quality of life, only gabapentin, pregabalin and duloxetine have been adequately studied and have shown positive effects, while effects of opioids are controversial.

Drugs with less established efficacy in various neuropathic conditions and recommended as second line therapy include lamotrigine, carbamazepine and the newer antidepressants venlafaxine and duloxetine (level A). However, venlafaxine and duloxetine may be considered as first line therapy in painful polyneuropathies.

Drugs with weak or controversial efficacy include capsaicin, SSRIs and mexiletine (level A), but capsaicin may be more effective on paroxysmal pain and mechanical allodynia (level B) and can be proposed if pain covers a limited area and if there is preservation of sensation. There is insufficient support for the use of oxcarbazepine (level C). In three RCTs topiramate was ineffective (level A). Combination therapy is recommended in case of insufficient efficacy with mono-therapy and should preferably use drugs with complementary mechanisms of action. It has been shown useful for gabapentin/morphine and gabapentin/venlafaxine (level A).

References

The references cited in the text can be found in:

1. Cruccu G, Anand P, Attal N, Garcia-Larrea L, Haanpa M, Jørum E, Serra J, Jensen TS. EFNS guidelines on neuropathic pain assessment. Eur J Neurol 2004; 11: 153–162
2. Attal N, Haanpaa M, Hansson P, Jensen TS, McQuay H, Nurmikko T, Sampaio C, Sindrup S, Wiffen P, Cruccu G. EFNS guidelines on medical treatment of neuropathic pain (in press)

Building the Evidence Base for Medical Devices: Strategies and Pitfalls

R. Taylor

Why the Increasing Need for Evidence?

There is an increasing global trend for healthcare policy makers to use evidence to assist their population level decisions, so-called evidence-based policy. Such policy decisions can be at the level of hospital, primary care trust, a region or a whole country [1]. In the current climate of rising healthcare costs, many healthcare providers and payers also wish to include not only consideration of evidence of clinical outcomes but also the costs of medical technologies, such as new medical devices or drugs. Evidence is therefore increasingly seen as a tool to assist health care policy makers in their efforts to contain costs.

This paper discusses a number of issues related to the evidence for medical devices and clinical related procedures: the changing face of evidence requirements of healthcare policy makers; the challenges of generating evidence for medical devices and procedures: and finally, some suggestions for how medical device evidence should be collected in the future.

What Do We Mean by "Evidence"?

For many years there has been a requirement for efficacy (i. e. what are the benefits of a medical device compared to no therapy or placebo) and safety data for a medical device for licensing and obtaining a CE mark in Europe or equivalent in other parts of the world. However, many European countries (e. g. UK, Sweden, Germany and the Netherlands) for a medical device to be listed or reimbursed by a healthcare payer have, in recent times, introduced the additional evidence hurdles of clinical effectiveness (i. e. what are the benefits of a medical device over and above current therapy?) and cost effectiveness (i. e. are these health benefits worth the additional cost of the medical device [2]?).

Superimposed on these evidential requirements, is the recognition for the hierarchy of evidence. According to the evidence hierarchy, the randomized controlled trial (RCT) is regarded as the highest level of evidence for judging the effectiveness of therapeutic interventions [3]. For example, evidence collected within two or more well conducted randomized-controlled trials would be regarded as "level I++ evidence" and as a result would receive a "grade A" policy recommendation. In contrast, evidence from an uncontrolled case series represents "level-IV evidence" and would receive a "grade D" policy recommendation. One frequently used example of such an evidence hierarchy [4] is shown in ◘ Table 4.1.

The Pitfalls of Building Evidence for Medical Devices

A number of characteristics of medical devices challenge the conduct of the classic double-blind random-

◘ Table 4.1. Levels of evidence and grades of policy-recommendations (adapted from [4])

	Levels of evidence
I++	High quality meta-analyses, systematic reviews of RCTs, or RCTs with a very low risk of bias
I+	Well-conducted meta-analyses, systematic reviews of RCTs, or RCTs with a low risk of bias
I	Meta-analyses, systematic reviews or RCTs, or RCTs with a high risk of bias
II++	High quality systematic reviews of case-control or cohort studies or high quality case-control or cohort studies with a very low risk of confounding, bias, or chance and a high probability that the relationship is causal
II+	Well conducted case-control or cohort studies with a low risk of confounding, bias, or chance and a moderate probability that the relationship is causal
II	Case-control or cohort studies with a high risk of confounding, bias, or chance and a significant risk that the relationship is not causal
III	Non-analytic studies, e.g. case reports, case series
IV	Expert opinion
	Grades of recommendations
A	At least one meta-analysis, systematic review, or RCT rated as I++ and directly applicable to the target population or a systematic review of RCTs or a body of evidence consisting principally of studies rated as I+ directly applicable to the target population and demonstrating overall consistency of results
B	A body of evidence including studies rated as II++ directly applicable to the target population and demonstrating overall consistency of results or extrapolated evidence from studies rated as I++ or I+
C	A body of evidence including studies rated as II+ directly applicable to the target population and demonstrating overall consistency of results or extrapolated evidence from studies rated as II++
D	Evidence level III or IV or extrapolated evidence from studies rated as II+

ized controlled trial, often undertaken for pharmaceuticals. Some of these key differences between medical devices and drugs that underpin these challenges are summarized in ◘ Table 4.2 below. The difficulty in undertaking medical device and clinical procedure trials has been comprehensively reviewed elsewhere [5, 6].

Nevertheless, as will be argued below, many of these difficulties of medical device clinical trial design can be overcome (at least in part) by innovative trial methodology and design.

Strategies for Building an Evidence Base

A case example is used to illustrate the strategies of building an evidence base: spinal cord stimulation (SCS) for patients with failed backed surgery syndrome/chronic leg and back pain (FBSS/CLBP).

Knowing Your Evidence Base

The first component of the strategy for evidence building is to thoroughly know the evidence base for the medical device. Systematic review and meta-analysis are recognized methods to comprehensively and explicitly assess the evidence base for a given

◘ Table 4.2. Differences between medical devices and drugs

Drugs	Devices
Unchanging compound	Constantly evolving
Complications increase with use	Complications decrease with use
Results unrelated to physician skill	Results vary with operator
Placebo usually available	Less placebo
Crossover rare	Crossover common

therapy [7] A systematic review of SCS for FBSS and CLBP identified 72 case series, one RCT and one co-hort study [8]. The RCT showed that patients receiving SCS experienced both a significantly higher level of pain relief and lower requirement for opiate analgesia compared to re-operation [9]. Using the Harbour and Miller evidence assessment scale, there is "level I+" evidence for the use of SCS in FBSS. However, despite level-I evidence, in Europe current management practice for FBSS patients would be more likely to be optimal non-surgical medical care rather than re-operation. Thus the policy recommendation was "grade B".

A systematic review provides the means by which evidence grading can be undertaken and also evidence gaps can be identified.

Moving up the Evidence Hierarchy

The second component of the strategy of building is to move as high as possible up the evidence hierarchy. In the case of SCS for FBSS, this required a new RCT to be designed and undertaken that compared SCS with conventional medical management. With the support of one of the device manufacturers (Medtronic Sarl), in 2000 a group of clinicians experienced in SCS and clinical trial specialists met to consider how best to design such a trial. As a result, the PROCESS (A Prospective, Randomized, Controlled, Multicenter Study to Evaluate the Effectiveness and Cost-Effectiveness of Spinal Cord Stimulation) trial was born [10].

◘ Table 4.3. Challenges of medical device trials

Characteristic	Challenge	Trial design solutions	PROCESS trial features
Device-specific issues			
Clinician/surgeon preference for device	Randomization	1. Offer randomization before referral to device center 2. Select clinically appropriate comparator 3. Offer all patients opportunity to receive device	SCS patients randomized to SCS or (non-surgical) conventional medical management with option for patients to cross over at 6 months
Interaction between device and clinician/surgeon	Learning curve	1. Recruit centers with experienced clinicians 2. Collect data over time and adjust for learning curve 3. Cluster randomization (randomize centers)	Recruited centers with experienced SCS implanters
Incremental change in device over time	Timing of trial	1. Head to head comparisons of device versions 2. Assess at device development milestones	Evaluation of SYNERGY SCS system
Invasive/semi-invasive procedure	Blinding of patients, clinicians and researchers	1. Sham procedure 2. ON/OFF design 3. Independent outcome assessment	Due to paresthesia effect of SCS, blinding was not possible
Non device-specific issues			
Refractory population	Finite sample size and lack of statistical power	Multicenter recruitment Power trial on primary outcome	Recruited SCS patients across x international centers
No accepted comparator	Non-standardization	Pragmatic trial design where centers apply local "best care"	Conventional medical management applied according to the practice of centers
Therapy-specific outcomes	Irrelevant outcomes for policy making	Collection of patient-focused outcomes	Outcomes included functional (Oswestry) and health-related quality of life outcomes (SF-36 and EQ-5D)

PROCESS was designed to both overcome a number of the potential challenges of medical device trials and also explicitly cater for the changing evidence needs of policy makers. The challenges of device trials and some of the solutions are summarized in ◘ Table 4.3.

Conclusions

There has traditionally been a requirement for efficacy and safety evidence for the licensing of medical devices. However, health-policy makers are increasingly expecting data of the "real world" clinical effectiveness and cost effectiveness of medical devices. Furthermore, such data needs to be collected using a randomized controlled trial design. The collection of such data requires innovative design of medical device clinical trials.

References

1. Muir Gray JA (ed) Evidence-based policy. Churchill Livingstone, Edinburgh, 1997
2. Taylor RS, Drummond MF, Salkeld G, Sullivan SD. Inclusion of cost effectiveness in licensing requirements of new drugs: the fourth hurdle. BMJ 2004; 329: 972–975
3. Sackett DL, Hayes RB, Guyatt GH, Tugwell P (eds) Clinical epidemiology: A basic science for clinical medicine, 2nd edn. Little, Brown and Company, Boston, 1991
4. Harbour R, Miller J. A new system for grading recommendations in evidence based guidelines. BMJ 2001; 323: 334–336
5. Seibert M et al. On behalf of EUCOMED. Health Technology Assessment for Devices in Europe. Int J Technol Healthcare 2002; 18: 734–740
6. Hartling L et al. Challenges of systematic reviews of clinical devices and procedures. Ann Intern Med 2005; 204: 1100–1111
7. Oxman AD, Cook DJ, Guyatt GH. Users' guides to the medical literature. VI. How to use an overview. Evidence-Based Medicine Working Group. JAMA 1994; 272: 1367–1371
8. Taylor RS, Taylor RS, Buscher E, Van Buyten J. Systematic review and meta-analysis of the effectiveness of spinal cord stimulation in the management of failed back surgery syndrome. Spine 2005; 30: 152–160
9. North RB, Kidd DH, Farrokhi F, Piantadosi S. Spinal cord stimulation versus repeated lumbosacral spine surgery for chronic pain: A randomized, controlled trial. Neurosurgery 2005; 56: 98–107
10. Kumar K, North R, Taylor R, Van den Abeele C, Gehring M et al. Spinal cord stimulation versus conventional medical management: a prospective, randomised, controlled, multicentre study of patients with Failed Back Surgery Syndrome (PROCESS study). Neuromodulation 2005; 8: 213–218

Part II:

Basic Research and Future Directions in Interventional Neuroscience

Mechanisms of Spinal Cord Stimulation in Neuropathic and Vasculopathic Pain: Present Status of Knowledge – and Views for the Future

B. Linderoth

Background

Spinal cord stimulation (SCS) emerged as a direct clinical spin-off from the gate control theory by Melzack and Wall in 1965 [34]. It has been estimated that, presently, more than 22,000 SCS systems are implanted every year worldwide and out of these more than 14,000 new cases. Although the physiological mechanisms behind the beneficial effects of this therapy, as yet, are only fragmentarily understood, the spread of the method is high. Paradoxically, and in contrast to predictions from the gate theory, SCS proved inefficacious in acute nociceptive pain conditions, and neuropathic pain of peripheral origin eventually emerged as the cardinal indication for this mode of treatment [22, 29, 32, 35, 37].

However, during the 80s reports demonstrated that SCS could alleviate also some types of nociceptive pain, i. e. selected ischemic pain states in e. g. peripheral arterial occlusive disease (PAOD), in vasospastic conditions and in therapy-resistant angina pectoris.

The exact mechanisms of action for SCS are thus still not mapped and only over the last few years more solid evidence of the underlying physiological mechanisms has emerged. Present concepts concerning the mechanisms of pain relief with SCS differ fundamentally between the use of this therapy in neuropathic and in ischemic/vasculopathic pain conditions (e. g. [29]).

Neurogenic Pain

In neuropathic pain the hyperexcitability demonstrated by multimodal wide-dynamic range (WDR) cells in the dorsal horns [49] seems to be related to increased basal release of excitatory amino acids e. g. glutamate, and a dysfunction of the local spinal GABA system [8, 42]. SCS has, in experiments on animal models of neuropathy, been demonstrated to inhibit dorsal horn (DH) WDR hyperexcitability and to induce release of GABA in the DHs, with a subsequent decrease of the interstitial glutamate concentration [8, 49]. The GABA release was solely observed in animals responding to SCS with symptom alleviation [42]. Activation of the GABA-B receptor seems to play a pivotal role for the suppression of glutamate release. Available evidence indicates that stimulation-induced release of adenosine [8], serotonin and noradrenalin [23, 25] (the two latter involved in descending inhibition) in the DH also may contribute (see ◘ Fig. 5.1 which schematically illustrates the circuitry involved in the effects of SCS). As a matter of fact, descending inhibition via a brain stem loop has been proposed as the principal mechanisms by some research groups (e. g. [12]).

In contrast, our own studies have so far indicated that only a minor part (<10%) of the DH inhibition is relayed by a supra-spinal loop [48].

However, a cascade release of neuroactive substances is probably induced by SCS both in the DHs

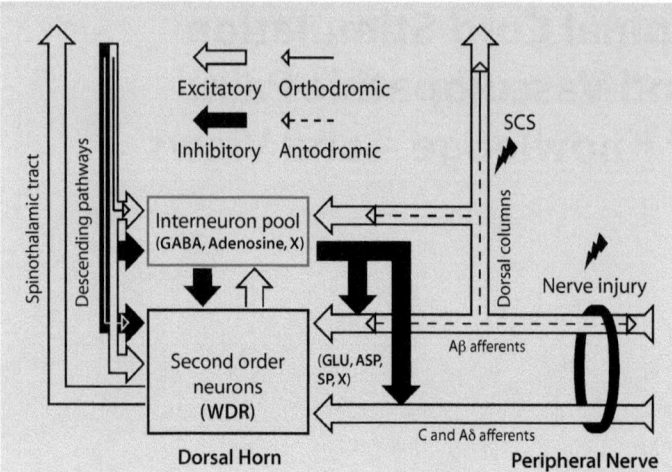

Fig. 5.1. Antidromic activation of dorsal columns is, via collaterals, mediated into the DHs, establishing contact with a multitude of neurons; among those GABAergic interneurons. A stimulation-induced release of GABA, binding to GABA-B receptors decrease release of EAAs, especially glutamate. However, many neuronal circuits take part in the inhibitions of the sensitized DH neurons, e. g. adenosinergic, serotonergic and no-radrenergic neurons, the two latter mediating descending inhibitory influence from supra-spinal centers. The major part of this circuitry is certainly, as yet, unknown (x). (Redrawn after Meyerson and Linderoth 2003)

and in other sites, e. g. in the brain stem [42], and multiple, as yet unknown, mechanisms thereby activated (review cf [31, 38]).

Peripheral Vasculopathic Pain

As already stated above, it is by now widely recognized that SCS does not alleviate acute nociceptive pain. However, in ischemic extremity pain, which is mainly nociceptive, SCS-induced relief of tissue ischemia seems to be the primary event either by increasing/re-distributing blood flow to the ischemic tissues (reviews in [23, 27, 31], or by decreasing tissue oxygen demand.

In PAOD, experimental studies favor the notion that SCS induces peripheral vasodilatation by suppressing efferent sympathetic activity resulting in diminished peripheral vasoconstriction and secondary relief of pain [24, 25, 27, 29], but recent evidence indicates that also antidromic mechanisms may be activated by SCS intensities far below the motor threshold and that this may result in peripheral CGRP release with subsequent peripheral vasodilatation [6, 30, 43, 45] (see ◘ Fig. 5.2 which summarizes the demonstrated dual-mechanism concept).

Recent animal studies have demonstrated that, which mechanism dominates, is related to the activity level of the sympathetic system – and possibly also to genetic and dietary differences [44].

Angina Pectoris

For coronary ischemia, manifesting as angina pectoris, the situation is also unclear. Although early animal data demonstrated direct inhibitory effects of SCS on cardiac nociception [5], it has later been clearly proven in clinical studies that SCS does not merely cause a pain blockade; resolution of cardiac ischemia remains the primary factor [33]. Some researchers favor a stimulation-induced flow increase or redistribution of blood supply (e. g. [16]), while others interpret the reduction of coronary ischemia (decreased ST changes; reversal of lactate production) as mainly due to decreased cardiomyocyte oxygen demand ([33]; review cf [11]). Experimental studies have hitherto been unable to demonstrate a local flow increase or redistribution of blood in the myocardium by SCS ([19]), but instead pre-emptive SCS seems to induce protective changes in the myocardium making it more resistant to critical ischemia (e. g. [4, 5]. Recent studies indicate that SCS-induced local catecholamine release in the myocardium [1] could trigger protective changes in the cardiomyocytes related to mechanisms behind "ischemic pre-conditioning".

Furthermore, SCS seems to exert arrhythmia control in the heart. In ischemia the intrinsic cardiac nervous system is profoundly activated. If this activity persists it may result in spreading dysrhythmias leading to more generalized ischemia. SCS stabilizes activity of these intrinsic ganglia especially at ischemic challenge and may in this way protect the heart from

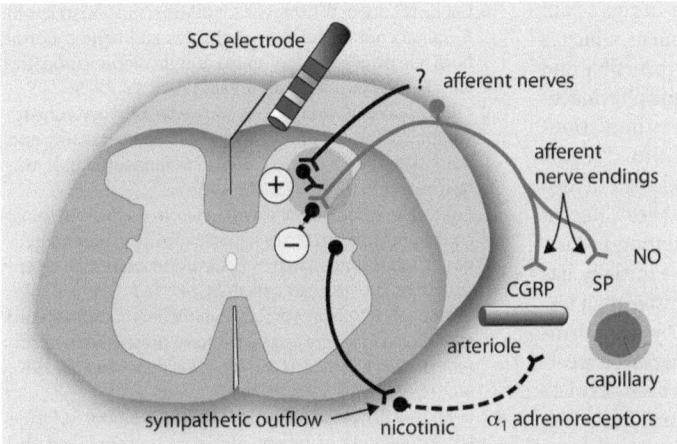

Fig 5.2. In alleviation of ischemic pain reduction of ischemia is the primary event. Also here multiple mechanisms seem to participate. Antidromic activation on hitherto unknown neuronal circuits reducing sympathetic outflow is one mechanism. The auto-nomic efference in question seems to have nicotinic ganglionic receptors and mainly alpha$_1$-adrenoreceptors at the neuro-effector junction. Another mechanism active at even low SCS intensities is the antidromic vasodilatation via activation of primary efferent fibers leading to peripheral release of CGRP with subsequent vasodilatation. The exact circuitry is not established but the presence of NO is required. Which mecha-nism that dominates seems to depend on the activity level of the sympathetic system. At low sympathetic tonus the antidromic activation dominates, but at higher levels, especially the later phase of vasodilatation seems to depend on sympathetic inhibi-tion [23]

more severe ischemic threats due to generalized ar-rhythmia [2]. The exact mechanisms remain to be discovered.

Pain with Dysautonomia

SCS has during the last years demonstrated its effi-cacy in complex regional pain syndromes (CRPS) [15, 18, 20, 41].

In principle SCS may act on the symptoms of CRPS in several ways:
- by a direct inhibitory action onto central hyper-excitable central neuronal circuits (as indicated above);
- by decreasing sympathetic efferent output acting on the de-novo activated adrenoreceptors on the damaged sensory neurons [23], and/or
- by reducing peripheral ischemia both by the anti-sympathetic action but also by e. g. antidromic mechanisms resulting in peripheral release of vasoactive substances e. g. CGRP and substance P ([6], cf also [3, 39, 47].

This 3rd action is related to the "indirect-coupling hypothesis" for dysautonomic pain conditions where the damaged afferent neurons develop hypersensitiv-ity to even mild hypoxia [39] lowering their activation thresholds. However, there are as yet no firm data to support these hypotheses and some recent clinical observations actually question the relevance of the "indirect-coupling mechanism" for the effect of SCS on pain in CRPS (e. g. [17]), but also conflicting data exist [15].

Conclusions and Future Directions

Thus, SCS induces effects in multiple systems and the benefit for a certain condition may depend on a selec-tion out of this cascade of biological changes. Knowl-edge about physiological mechanisms behind the beneficial effects provides a corner-stone for further development of stimulation methods and future strategies to support these techniques with e g. ad-ministration of receptor-active pharmaceuticals in cases with inadequate response to stimulation per se (cf [21, 40, 46]. SCS is a therapy, effective in some pain syndromes otherwise resistant to treatment, which is lenient to patients, minimally invasive, reversible and with few side-effects compared to chronic pharmaco-therapy. One may, in fact regard the stimulation-induced changes as effects of an in situ ("super-targeted") delivery of physiologically relevant neuro-active substances in "physiological dosages". Recent research also demonstrates that the method is cost-beneficial, and although more expensive initially, has in several studies proved to be less expensive than tra-ditional treatment after in average 2.5–3 years. I firm-ly believe that SCS at present is an under-used treat-ment modality which unfortunately has been very late in systematic exploration and comparison of different stimulation modes, paradigms and over-all regimens. The further exploration of the mechanisms of action and of the most effective modes of stimulation are the two areas which at present demand the highest priori-ties. The technical developments needed are already underway enabled by the progress in electronics and forced by the increased commercial competition.

References

1. Ardell JL, Southerland EM, Milhorn D, Foreman RD, Linderoth B, DeJongste MJL, Armour JA. Pre-emptive spinal cord stimulation mitigates myocardial ischemia-induced infarction via alpha adrenergic receptors. Exp Biol 2005; 2–6, Abstract 356.10

2. Armour JA, Linderoth B, Arora RC, DeJongste MJ, Ardell JL, Kingma JG Jr, Hill M, Foreman RD. Long-.term modulation of the intrinsic cardiac nervous system by spinal cord neu-rons in normal and ischaemic hearts. Auton Neurosci 2002; 95: 71–79

3. Baron R, Levine JD, Fields HL. Causalgia and reflex sympa-thetic dystrophy: does the sympathetic nervous system contribute to the generation of pain? Muscle Nerve 1999; 22: 678–695

4. Cardinal R, Ardell JL, Linderoth B, Vermeulen M, Foreman RD, Armour JA. Spinal cord activation differentially modulates ischaemic electrical responses to different stressors in ca-nine ventricles. Auton Neurosci 2004; 111: 37–47

5. Chandler MJ, Brennan TJ, Garrison DW, Kim KS, Schwartz PJ, Foreman RD. A mechanism of cardiac pain suppression by spinal cord stimulation: implications for patients with an-gina pectoris. Eur Heart J 1993; 14: 96–105

6. Croom JE, Foreman RD, Chandler MJ, Barron KW. Re-eval-uation of the role of the sympathetic nervous system in cutaneous vasodilation during dorsal spinal cord stimula-tion: are multiple mechanisms active? Neuromodulation 1998; 1: 91–101

7. Cui J-G, Linderoth B, Meyerson BA: Effects of spinal cord stimulation on touch-evoked allodynia involve GABAergic mechanisms. An experimental study in the mononeuro-pathic rat. Pain 1996; 66: 287–295

8. Cui JG, O'Connor WT, Ungerstedt U, Meyerson BA, Linderoth B. Spinal cord stimulation attenuates augmented dorsal horn release of excitatory amino acids in mononeuropathy via a GABAergic mechanism. Pain 1997b; 73: 87–95

9. Cui JG, Sollevi A, Linderoth B, Meyerson BA. Adenosine re-ceptor activation suppresses tactile hypersensitivity and potentiates effect of spinal cord in mononeuropathic rats. Neurosci Lett 1997c; 223: 173–176

10. Cui JG, Meyerson BA, Sollevi A, Linderoth B. Effects of spinal cord stimulation on tactile hypersensitivity in mononeuro-pathic rats is potentiated by GABAB and adenosine recep-tor activation. Neurosci Lett 1998; 247: 183–186

11. Eliasson T, Augustinsson LE, Mannheimer C. Spinal cord stimulation in severe angina pectoris- presentation of cur-rent studies, indications and clinical experience. Pain 1995; 65: 169–179

12. El-Khoury C, Hawwa N, Baliki M, Atweh SF, Jabbur SJ, Saade NE. Attenuation of neuropathic pain by segmental and supraspinal activation of the dorsal column system in awake rats. Neuroscience 2002;112: 541–553

13. Foreman RD, DeJongste MJL, Linderoth B. Integrative con-trol of cardiac function by cervical and thoracic spinal neurons. In: Armour JA, Ardell JL (eds) Basic and clinical neurocardiology. Oxford University Press, London, 2004; pp 153–186

14. Foreman RD, Linderoth B, Ardell JL, Barron KW, Chandler MJ, Hull Jr. SS, TerHorst GJ, DeJongste MJL, Armour JA. Mod-ulation of intrinsic cardiac neurons by spinal cord stimula-tion: implications for therapeutic use in angina pectoris. Cardiovasc Res 2000; 47: 367–375

15. Harke H, Gretenkort P, Ladleif HU, Rahman S. Spinal cord stimulation in sympathetically maintained complex re-gional pain syndrome type I with severe disability. A pro-spective clinical study. Eur J Pain 2005; 9: 363–373

16. Hautvast RW, Blanksma PK, DeJongste MJ, Pruim J, van der Wall EE, Vaalburg W, Lie KI. Effect of spinal cord stimulation on myocardial blood flow assessed by positron emission tomography in patients with refractory angina pectoris. Am J Cardiol 1996; 77:462–467

17. Kemler MA, Barendse GA, van Kleef M, Egbrink MG. Pain relief in complex regional pain syndrome due to spinal cord stimulation does not depend on vasodilation. Anesthesiol-ogy 2000a; 92: 1653–1660

18. Kemler MA, Barendse GA, van Kleef M, de Vet HC, Rijks CP, Furnee CA, van den Wildenberg FA. Spinal cord stimulation in patients with chronic reflex sympathetic dystrophy. N Engl J Med 2000b; 343: 618–624

19. Kingma JG, Linderoth B, Ardell JL, Armour JA, DeJongste MJL, Foreman RD. Neuromodulation therapy does not influ-ence blood flow distribution or left ventricular dynamics during acute myocardial ischaemia. Auton Neurosci Basic Clin 2001; 91: 47–54

20. Kumar K, Nath RK, Tooth C. Spinal cord stimulation is effec-tive in the management of reflex sympathetic dystrophy. Neurosurgery 1997; 40: 736–747

21. Lind G, Meyerson BA, Winter J, Linderoth B. Intrathecal baclofen as adjuvant therapy to enhance the effect of spinal cord stimulation in neuropathic pain: a pilot study. Eur J Pain 2004; 8: 877–883

22. Lindblom U, Meyerson BA. Influence of touch, vibration and cutaneous pain of dorsal column stimulation in man. Pain 1975; 1: 257–270

23. Linderoth B, Foreman RD. Physiology of spinal cord stimulation. Review and Update. Neuromodulation 1999; 2: 150–164

24. Linderoth B, Fedorcsak I, Meyerson BA. Peripheral vasodilatation after spinal cord stimulation: animal studies of putative effector mechanisms. Neurosurgery 1991; 2: 187–195

25. Linderoth B, Gazelius B, Franck J, Brodin E. Dorsal column stimulation induces release of serotonin and substance P in the cat dorsal horn. Neurosurgery 1992; 31:289–297

26. Linderoth B, Gunasekera L, Meyerson BA. Effects of sympathectomy on skin and muscle microcirculation during dorsal column stimulation: animal studies. Neurosurgery 1991; 29: 874–879

27. Linderoth B, Herregodts P, Meyerson BA. Sympathetic mediation of peripheral vasodilation induced by spinal cord stimulation: animal studies of the role of cholinergic and adrenergic receptor subtypes. Neurosurgery 1994; 35: 711–719

28. Linderoth B. Spinal cord stimulation in ischemia and ischemic pain. In: Horsch S, Claeys L (eds) Spinal cord stimulation: an innovative method in the treatment of PVD and angina. Steinkopff Verlag, Darmstadt, 1995, pp 19–35

29. Linderoth B, Meyerson BA. Dorsal column stimulation: modulation of somatosensory and autonomic function. In: McMahon SB, Wall PD (eds) The neurobiology of pain. Seminars in the Neurosciences, vol 7. Academic Press, London, 1995, pp 263–277

30. Linderoth B, Foreman RD. Physiology of spinal cord stimulation. Review and Update. Neuromodulation 1999; 2: 150–164

31. Linderoth B, Meyerson BA. Spinal cord stimulation: I: Mechanisms of action. In: Burchiel KJ (ed) Surgical management of pain. Thieme, Stuttgart New York; 2002, pp 505–526

32. Linderoth B, Simpson B, Meyerson BA. Spinal cord and brain stimulation. In: McMahon S, Kolzenburg M (eds) Wall and Melzack's textbook of pain, 5th edn. Elsevier, Amsterdam; 2005

33. Mannheimer C, Eliasson T, Andersson B, Bergh CH, Augustinsson LE, Emanuelsson H, Waagstein F. Effects of spinal cord stimulation in angina pectoris induced by pacing and possible mechanism of action. Br Med J 1993; 307: 477–480

34. Melzack R, Wall PD. Pain mechanisms: a new theory. Science 1965; 150: 971–979

35. Meyerson BA, Linderoth B. Therapeutic electrical neurostimulation from a historical perspective. In: Merskey H, Loeser JD, Dubner R (eds) The paths of pain 1975–2005. IASP Press, Seattle; 2005, pp 313–327

36. Meyerson BA, Linderoth B. Mechanisms of spinal cord stimulation in neuropathic pain. Invited review. Neurol Res 2000; 22: 285–292

37. Meyerson BA, Linderoth B. Spinal cord stimulation. In: Loeser JD (ed) Bonica's management of pain, 3rd edn. Lippincott Williams and Wilkins, Philadelphia; 2000, pp 1857–1876

38. Meyerson BA, Linderoth B. Spinal cord Stimulation: mechanisms of action in neuropathic and ischemic pain. In: Simpson BA (ed) Electrical stimulation and the relief of pain, vol 15. Elsevier, New York; 2003, pp 161–182

39. Michaelis M. Coupling of sympathetic and somatosensory neurons following nerve injury: Mechanisms and potential significance for the generation of pain. In: Devor M, Rowbotham MC, Wiesenfeld Z (eds) Progress in pain research and management, vol 16. IASP Press, Seattle; 2000; pp 645–656

40. Schechtmann G, Wallin J, Meyerson BA, Linderoth B. Intrathecal clonidine potentiates suppression of tactile hypersensitivity by spinal cord stimulation in a model of neuropathy. Anesth Analg 2004; 99: 135–139

41. Stanton-Hicks M. Spinal cord stimulation in the management of complex regional pain syndromes. Neuromodulation 1999; 2: 193–201

42. Stiller CO, Cui J-G, O'Connor WT, Brodin E, Meyerson BA, Linderoth B. Release of GABA in the dorsal horn and suppression of tactile allodynia by spinal cord stimulation in mononeuropathic rats. Neurosurgery 1996; 39: 367–375

43. Tanaka S, Barron KW, Chandler MJ, Linderoth B, Foreman RD. Low intensity spinal cord stimulation may induce cutaneous vasodilation via CGRP release. Brain Res 2001; 896: 183–187

44. Tanaka S, Barron KW, Chandler MJ, Linderoth B, Foreman RD. Local cooling alters neural mechanisms producing changes in peripheral blood flow by spinal cord stimulation. Autonomic Neuroscience: Basic and Clinical 2003; 104: 117–127

45. Tanaka S, Komori N, Barron KW, Chandler MJ, Linderoth B, Foreman RD. Mechanisms of sustained cutaneous vasodilation induced by spinal cord stimulation. Auton Neurosci 2004; 114: 55–60

46. Wallin J, Cui J-G, Yakhnitsa V, Meyerson BA, Linderoth B. Gabapentin and Pregabalin suppress tactile allodynia and potentiate spinal cord stimulation in a model of neuropathy. Europ J Pain 2002; 6: 261–272

47. Wasner G, Schattschneider J, Heckmann K, Maier C, Baron R. Vascular abnormalities in reflex sympathetic dystrophy (CRPS I): Mechanisms and diagnostic value. Brain 2001; 124: 587–599

48. Yakhnitsa V, Linderoth B, Meyerson BA. Modulation of dorsal horn neuronal activity by spinal cord stimulation in a rat model of neuropathy: the role of the dorsal funicles. Neurophysiology 1998; 30: 424–427

49 Yakhnitsa V, Linderoth B, Meyerson BA. Spinal cord stimulation attenuates dorsal horn neuronal hyperexcitability in a rat model of mononeuropathy. Pain 1999; 79: 223–233

Objective Measurement of Physical Activity in Patients with Chronic Lower Limb Pain Treated with Spinal Cord Stimulation

E. Buchser, A. Paraschiv-Ionescu, A. Durrer, K. Aminian

Introduction

Spinal cord stimulation (SCS) is increasingly used for the treatment of intractable pain syndromes due to vascular [1] or neurogenic disorders [2]. Neuromodulation therapies, whatever the indication, are symptomatic treatments and their benefits are therefore measured in terms of quality of life (QOL) rather than disease cure. One of the main features of QOL is the ability to perform unimpaired physical activity. Any condition that results in the decrease of physical activity is associated with a decrease in QOL, and higher levels of activity are associated with improved well-being [3]. A number of factors can affect the physical performance including age, cardio-respiratory fitness, body weight [4], fear of pain [5–7] and belief about activity [8]. It is generally accepted that chronic pain states are associated with a decreased level of physical activity, although there seems to be no direct correlation between the intensity of pain and the restriction of physical activity [9].

Quality of life is multidimensional [10] and is usually measured with validated questionnaires [11–16] that assess physical, material, social and emotional well-being as well as physical activity [17].

Contrary to the emotions that escape objective assessment, the physical activity can be measured in quantitative terms using a variety of techniques. The level of physical activity can be a crucial part of the evaluation of the state of a disease or the efficacy of a treatment. Since it has long been recognized that self-reporting is unreliable [18, 19], objective measurements of the quality and the quantity of physical activity can be important.

The research into the assessment of physical activity has been driven mostly by queries regarding the energy expenditure [20] and the analysis of the gait, either in neurological and orthopedic disorders or to characterize the risk of falling, particularly in the elderly [21]. Techniques that are not restricted to the environment of a laboratory and therefore offer the possibility to carry out measurement in normal life conditions, have undergone substantial sophistication. Over the last two decades they have evolved from foot switches and pedometers [22, 23] to the use of sophisticated kinematical sensors combining accelerometers and gyroscopes [24–26].

We have used a validated method [27] to study the quantitative and qualitative changes in the physical activity and the gait parameters in a group of chronic pain patients treated with spinal cord stimulation.

Summary of Background Data and Methods

Using 3 kinematical sensors attached to the chest, the thigh and the calf we measured the time spent in different body postures (lying, sitting, standing), the walking activity and the gait parameters. Changes in the spontaneous physical activity following SCS were evaluated under real life conditions. Five series of measurements (carried out each during 7 hours per day on 5 consecutive days) were performed before the implant and at one, three, six and 12 months after the implantation of a SCS system. At the beginning of each series of measurements the least and the usual pain intensity scores were obtained in all patients using a 10 cm visual analogue scale (VAS).

Preliminary results from an ongoing study have been recently published [28]. All patients suffered from chronic intractable pain caused by spinal stenosis, peripheral artery disease or peripheral neuropathy, which resulted in a decreased walking perimeter.

Results

Compared to baseline values, the average VAS score (VAS 59 mm ± 17), decreased significantly (p<0.05) at all times after the implantation of a SCS system 61%, 60%, 51%, and 81% at one, three, six, and 12 months (VAS 10 mm ± 9), respectively. The physical activity increased consistently during the entire follow-up period and the time spent lying (600 min + 392 at baseline) decreased steadily and was significantly lower at six (p<0.001), and 12 months (197 min ± 237, p<0.01). The average total walking distance (3064 m ± 2797 at baseline) increased by 146%, 206%, 328% and 389% at 1, 3, 6 and 12 months (6478 m ± 2914) respectively, reaching statistical significance (p<0.05) after 1 month. The gait parameters improved progressively. The stride length and the speed were significantly increased (p<0.05) at all times. The variability between patients decreased progressively over the follow-up period. The trend was similar in all patients and for all parameters.

Conclusions and Implications for Clinical Practice Today

Meaningful assessment of the physical activity of daily living must be carried out over a prolonged period of time and should include an analysis of the gait parameters. In a study looking at how patients validate the outcomes that are commonly used in the evaluation of SCS, the majority of patients reported that the inability to walk "normal" distances without pain was the predominant factor [29]. Yet most studies failed to include objective assessments of physical activity. Current evidence suggests that SCS-related

pain relief is associated with a significant improvement of the physical activity that is consistent, progressive and appears to be maintained over time.

Since the ability to perform physical activity is closely correlated with the QOL, we suggest that quantitative and reliable assessment of the spontaneous physical activity may be useful to further establish the beneficial effect of symptomatic treatment in general and SCS for the management of pain in particular.

Future Directions

It is well recognized that under healthy conditions, most physiological signals have rhythms. Earlier studies showed that with normal gait, the stride interval display subtle and complex fluctuations consistent with a fractal rhythm organization. This organization can be disrupted in disease states to a degree that is related to the functional impairment [30].

Modern theories of nonlinear complex system can be applied to analyze time series of long-term recordings of physical activity (postures and gait parameters as described above). Using the detrended fluctuation analysis we found that chronic pain is associated with fractal pattern disruption of the daily life's physical activity, which tends to be restored with pain relief.

Given the unspecific and ubiquitous occurrence of the fractal rhythms organization of biological processes, the disruption of fractality in physical activity parameters may be linked to similar alterations of other functions (such as heart rate, blood pressure, breathing, ion channel kinetics), which may lead to fundamentally new approaches in the understanding, the diagnosis, the prevention or the treatment of a number of disorders.

References

1. Ubbink DT, Jacobs MJ. Spinal cord stimulation in critical limb ischemia. A review. Acta Chir.Belg 2000; 100: 48–53 [2001; 100: 48–53]
2. Linderoth B, Meyerson B. Peripheral and central nervous system stimulation in chronic therapy-resistant pain. Background, hypothetical mechanisms and clinical experiences. Lakartidningen 2001; 98: 5328–5334, 5336
3. Stewart AL, Hays RD, Wells KB, Rogers WH, Spritzer KL, Greenfield S. Long-term functioning and well-being outcomes associated with physical activity and exercise in patients with chronic conditions in the Medical Outcomes Study. J Clin Epidemiol 1994; 47: 719–730

4. Coakley EH, Kawachi I, Manson JE, Speizer FE, Willet WC, Colditz GA. Lower levels of physical functioning are associated with higher body weight among middle-aged and older women. Int J Obes Relat Metab Disord 1998; 22: 958–965

5. Crombez G, Vlaeyen JW, Heuts PH, Lysens R. Pain-related fear is more disabling than pain itself: evidence on the role of pain-related fear in chronic back pain disability. Pain 1999; 80: 329–339

6. Vowles KE, Gross RT. Work-related beliefs about injury and physical capability for work in individuals with chronic pain. Pain 2003; 101: 291–298

7. Vlaeyen JW, de Jong J, Geilen M, Heuts PH, van Breukelen G. The treatment of fear of movement/(re)injury in chronic low back pain: further evidence on the effectiveness of exposure in vivo. Clin J Pain 2002; 18: 251–261

8. Silver A, Haeney M, Vijayadurai P, Wilks D, Pattrick M, Main CJ. The role of fear of physical movement and activity in chronic fatigue syndrome. J Psychosom Res 2002; 52: 485–493

9. Fordyce WE, Lansky D, Calsyn DA, Shelton JL, Stolov WC, Rock DL. Pain measurement and pain behavior. Pain 1984; 18: 53–69

10. Felce D, Perry J. Quality of life: its definition and measurement. Res Dev Disabil 1995; 16: 51–74

11. Melzack R. The McGill Pain Questionnaire: major properties and scoring methods. Pain 1975; 1: 277–299

12. Holroyd KA, Holm JE, Keefe FJ, Turner JA, Bradley LA, Murphy WD, Johnson P, Anderson K, Hinkle AL, O'Malley WB. A multicenter evaluation of the McGill Pain Questionnaire: results from more than 1700 chronic pain patients. Pain 1992; 48: 301–311

13. Brazier J. The Short-Form 36 (SF-36) Health Survey and its use in pharmacoeconomic evaluation. Pharmacoeconomics 1995; 7: 403–415

14. Essink-Bot ML, Krabbe PF, Bonsel GJ, Aaronson NK. An empirical comparison of four generic health status measures. The Nottingham Health Profile, the Medical Outcomes Study 36-item Short-Form Health Survey, the COOP/WONCA charts, and the EuroQol instrument. Med Care 1997; 35: 522–537

15. Brazier JE, Walters SJ, Nicholl JP, Kohler B. Using the SF-36 and Euroqol on an elderly population. Qual Life Res 1996; 5: 195–204

16. Schroll M, Schlettwein D, van Staveren W, Schlienger JL. Health related quality of life and physical performance. SENECA 1999. J Nutr Health Aging 2002; 6:15–19

17. EuroQol – a new facility for the measurement of health-related quality of life. The EuroQol Group. Health Policy 1990; 16:199–208

18. Bassett DR, Cureton AL, Ainsworth BE. Measurement of daily walking distance-questionnaire versus pedometer. Med Sci Sports Exercise 2000; 32: 1018–1023

19. Kremer EF, Block A, Gaylor MS. Behavioral approaches to treatment of chronic pain: the inaccuracy of patient self-report measures. Arch Phys Med Rehabil 1981; 62: 188–191

20. Schutz Y, Weinsier RL, Hunter GR. Assessment of free-living physical activity in humans: an overview of currently available and proposed new measures. Obes Res 2001; 9: 368–379

21. Najafi B. Physical activity monitoring and the risk of falling evaluation in elderly people. PhD Dissertation No 2672 (2002), Ecole Polytechnique Fédérale de Lausanne, 2003

22. Holden J, Fernie GR, Soto M. An assessment of a system to monitor the activity of patients in a rehabilitation programme. Prosthet Orthot Int 1979; 3: 99–102

23. Bassey EJ, Dallosso HM, Fentem PH, Irving JM, Patrick JM. Validation of a simple mechanical accelerometer (pedometer) for the estimation of walking activity. Eur J Appl Physiol Occup Physiol 1987; 56: 323–330

24. Aminian K, Najafi B, Bula C, Leyvraz PF, Robert P. Spatio-temporal parameters of gait measured by an ambulatory system using miniature gyroscopes. J Biomech 2002; 35: 689–699

25. Najafi B, Aminian K, Paraschiv-Ionescu A, Loew F, Bula CJ, Robert P. Ambulatory system for human motion analysis using a kinematic sensor: Monitoring of daily physical activity in the elderly. IEEE Transactions on Biomedical Engineering 2003; 50: 711–723

26. Aminian K, Robert P, Buchser EE, Rutschmann B, Hayoz D, Depairon M. Physical activity monitoring based on accelerometry: validation and comparison with video observation. Med Biol Eng Comput 1999; 37: 304–308

27. Paraschiv-Ionescu A, Buchser E, Rutschmann B, Najafi B, Aminian K. Ambulatory system for the quantitative and qualitative analysis of gait and posture in chronic pain patients treated with spinal cord stimulation. Gait Posture 2004; 20: 113–125

28. Buchser E, Paraschiv-Ionescu A, Durrer A, Depierraz B, Aminian K, Najafi B, Rutschmann B. Improved physical activity in patients treated for chronic pain by spinal cord stimulation. Neuromodulation 2005; 8: 40–48

29. Anderson VC, Carlson C, Shatin D. Outcomes of spinal cord stimulation: Patient validation. Neuromodulation 2001; 4: 11–17

30. Goldberger AL, Amaral LA, Hausdorff JM, Ivanov P, Peng CK, Stanley HE. Fractal dynamics in physiology: alterations with disease and aging. Proc Natl Acad Sci USA 2002; 99 (Suppl 1): 2466–2472

How Could HFS Functionally Inhibit Neuronal Networks?

A.-L. Benabid, S. Chabardes, E. Seigneuret, V. Fraix, P. Krack, P. Pollak

Main Clinical Effects of High Frequency Stimulation (HFS)

It has been observed incidentally in 1987, during a thalamotomy for essential tremor where electrical stimulation at low frequency was used to locate the target, that the tremor did not change significantly under stimulation at 30 to 50 Hertz, while it was stopped at 100 Hertz [1]. Since that time, the concept that HFS could induce a functional inhibition, and then could mimic a lesion, has been applied to all available targets for functional neurosurgery such as the thalamus and the pallidum. This concept was further extended to all the targets suggested by the basic research, such as the subthalamic nucleus (STN) reinvestigated in 1990 [5].

STN HFS has been used since 1993 to control the triad of symptoms in Parkinson's disease. It significantly decreases or even suppresses the akinesia, rigidity and tremor as well as off period dystonias. The quality of the results depends on the quality of the surgery but also on the quality of patient selection. The best candidates are Parkinsonian patients with idiopatic Parkinson's Disease (PD) who are levodopa responsive, at the stage of motor fluctuations and dyskinesias without neurocognitive alteration nor general contraindication.

It has been clearly shown [8] that there was a significant linear relationship between the quality of the improvement induced by HFS of STN and the quality of the improvement under the best levodopa treatment. In addition to this major target which constitutes currently the largest indication of HFS, over the last few years and particularly the last two years, a large number of other targets have been included in the field of HFS such as the nucleus accumbens, the internal capsule, the zona incerta, the radiatio prelemniscalis, the thalamus (in VIM, in CM-Pf, in the anterior nuclei), the STN, the internal pallidum (GPi), the ventro-medial hypothalamus and probably the lateral hypothalamus, the posterior hypothalamus, the PAG-PVG and recently the cortex, not to mention the posterior group of the thalamus which was stimulated at low frequency for pain.

The indications are currently more or less validated, but mostly subjected for those which are not validated, to multicenter clinical trials. The indications are in alphabetical order: aggressivity, cluster headache [12], depression [17], dystonia [9] and tardive dystonia, epilepsy [2, 7, 16], essential tremor, OCD [20], Parkinson's disease, TICs of the disease of Gilles de la Tourette [21].

For all those indications (except pain and deafferentation tremor which respond to low frequency stimulation), the effect is obtained at frequencies higher than 100 Hertz and mimics globally the effect of ablative lesions in the same areas, when this has been performed.

What Is the Mechanism?

This is the main question, which is currently not fully answered. The basic concepts are that electrical stimulation excites neural elements through membrane depolarization and that axons are excited at lower thresholds than cell bodies. There is therefore a paradoxical lesion-like effect of HFS and the question is how to solve this paradox between the fact that at high frequency, meaning more than 50 Hertz, mostly between 130 and 185 Hertz, HFS mimics the effect of ablative procedures considered of course as inhibitory in neural somatic structures (the thalamus, basal ganglia and the hypothalamus) while high frequency and low frequency stimulation both excite neural fiber bundles, as can be observed during surgery where HFS induces flashes in the optic track, contractions in the pyramidal track, paresthesias in the lemniscus medialis, and ocular deviations in the area of the third nerve fibers.

The mechanism can be considered currently as a complex association of various mechanisms, each of them affected with variable coefficients, and the effect might be more complex than it could be following only lesions. Lesions obviously destroy cells and fibers, both fibers outgoing from the target, and fibers passing through the target. On the contrary, HFS could be associating cell firing inhibition, excitation of inhibitory afferents, excitation of excitatory efferents in turn exciting inhibitory cells, exhaustion of neurotransmitters from the nuclei and cells which are being stimulated. The question is therefore to understand what happens and what are those respective weights for the different sub-mechanisms.

What Does the HFS of STN in the Basal Ganglia?

It has been shown in rats [4] that HFS of STN decreases the activity of this nucleus and in turn, by shutting down the glutamate output, decreases the activity in the entopedoncular cells (which is the equivalent of GPi in humans), as well as in the substantia nigra pars reticulata cells (SNpr) which are both GABAergic. Their hypo-activity would therefore induce the liberation of the substantia nigra compacta (SNpc) cells and all the ventral lateral group of the thalamus cells which are reactivated. As a consequence also, the reticularis thalami cells are inhibited as expected from the basic scheme of the network. This could be due to the retrograde excitation of the GP cells (which correspond to GPe in humans), however the lesion of this nucleus does not suppress the effect of HFS of STN.

Is Excitation the Mechanism?

This works at high frequency only. Could it be a temporal summation? This could be suggested by the fact that the curve is the same as the intensity-frequency curve shown in handbooks, based on the data obtained during the excitation of the fibers of the crab muscle nerve. This could also suggest that fibers are involved in this mechanism, but HFS works only in groups of neural somata, and not in fiber bundles: this would in turn suggest that cell bodies are involved.

Therefore, how could excitation and total inhibition coincide with a similar outcome in every structure tested so far?

Jamming could be a determining factor. At the neuronal level, there are several data showing that HFS decreases or strongly perturbs cell firing. This could be due to jamming or a feed-back loop, as we suggested particularly in the cases of tremor, where the jamming of the signal by HFS could render the system unable to process the information. This is what [14], has been called "zero variance" leading to a "functional lesion". This jamming is also visible in living monkeys or rat brain slices [13, 15] showing that jamming would perturb the transmission through the neural structure of wrong neural messages which are therefore becoming meaningless. Similar effects have been shown in slices of hippocampus [6].

The silencing of neuronal firing has been shown in the GPi [10], as well as in the STN [11, 22] of patients. It has been also shown [18] that immediately after stimulation, there is a strong decrease of activity, suggesting resetting, jamming or desynchronization.

What Happens at the Level of Passing Axons?

There is truly an excitation of these fibers and this could account for the side effects observed, particularly in the vicinity of STN, surrounded by extremely noisy neighbors which are responsible for the side effects seen when HFS involves the optic track, the pyramidal track, the lemnicus medialis, and third nerve fibers, in addition to the involvement of nuclei from the surrounding structures such as the thalamus, zona incerta and the substantia nigra pars reticulata.

What Happens at the Level of Efferent Axons?

Obviously, the axons exiting out of STN, are excited, according to the general observation that fibers are excited regardless of the frequency. Therefore, the excitation of those fibers during their course between the STN and the Gpi should induce an excitation at the subthalamo-pallidal synaptic level. We have performed studies using cell cultures where stimulation at high frequency significantly decreases the production of neurotransmitters, which would explain why excitation of these fibers might not necessarily induce an excitation at the post synaptic level (Xia et al., unpublished data). Similarly, cellular and molecular effects can be observed, using genomic and proteomic methods which are not described here.

Then one can produce a sketch of the putative multi-modality mechanism [3].

Conclusion

Do we have a model? High frequency and low frequency excite axons passing by the stimulation site while low frequency excites the cell bodies. HFS inhibits neuronal firing and may jam the neuronal messages, whether they are normal or wrong. Axons originating from the stimulated neurons convey spikes drifting to the cell bodies as well as to the synapses. HFS inhibits intra cellular protein synthesis processes. Synapses receiving spikes fire blank, devoid of neurotransmitters.

References

1. Benabid AL, Pollak P, Louveau A, Henry S, de Rougemont J. Combined (thalamotomy and stimulation) stereotactic surgery of the Vim thalamic nucleus for bilateral Parkinson disease. Appl Neurophysiol 1987; 50: 344–346

2. Benabid AL, Minotti L, Koudsie A, de Saint Martin A, Hirsch E. Antiepileptic effect of high-frequency stimulation of the subthalamic nucleus (corpus luysi) in a case of medically intractable epilepsy caused by focal dysplasia: a 30-month follow-up: technical case report. Neurosurgery 2002; 50: 1385–1391

3. Benabid AL, Wallace B, Mitrofanis J, Xia R, Piallat B, Chabardes S, Berger F. A putative generalized model of the effects and mechanism of action of high frequency electrical stimulation of the central nervous system. Acta Neurol Belg 2005; 105: 149–157

4. Benazzouz A, Piallat B, Pollak P, Benabid AL. Responses of substantia nigra reticulata and globus pallidus complex to high frequency stimulation of the subthalamic nucleus in rats: electrophysiological data. Neurosci Lett 1995; 189: 77–80

5. Bergman H, Wichmann T, DeLong MR. Reversal of experimental parkinsonism by lesions of the subthalamic nucleus. Science 1990; 249:1436–1438

6. Bikson M, Lian J, Hahn PJ, Stacey WC, Sciortino C, Durand DM. Suppression of epileptiform activity by high frequency sinusoidal fields in rat hippocampal slices. J Physiol 2001; 531:181–191

7. Chabardes S, Kahane P, Minotti L, Koudsie A, Hirsch E, Benabid AL. Deep brain stimulation in epilepsy with particular reference to the subthalamic nucleus. Epileptic Disord 2002; 4 (Suppl 3): S83–93

8. Charles PD, Van Blercom N, Krack P et al. Predictors of effective bilateral subthalamic nucleus stimulation for PD. Neurology 2002; 59: 932–934

9. Coubes P, Roubertie A, Vayssiere N, Hemm S, Echenne B. Treatment of DYT1-generalised dystonia by stimulation of the internal globus pallidus. Lancet 2000; 355: 2220–2221

10. Dostrovsky JO, Levy R, Wu JP, Hutchison WD, Tasker RR, Lozano AM. Microstimulation-induced inhibition of neuronal firing in human globus pallidus. J Neurophysiol 2000; 84: 570–574

11. Filali M, Hutchison WD, Palter VN, Lozano AM, Dostrovsky JO. Stimulation-induced inhibition of neuronal firing in human subthalamic nucleus. Exp Brain Res 2004; 156:274–281

12. Franzini A, Ferroli P, Leone M, Broggi G. Stimulation of the posterior hypothalamus for treatment of chronic intractable cluster headaches: first reported series. Neurosurgery 2003; 52:1095–1099; discussion 1099–1101

13. Garcia L, Audin J, D'Alessandro G, Bioulac B, Hammond C. Dual effect of high-frequency stimulation on subthalamic neuron activity. J Neurosci 2003; 23:8743–8751

14. Grill WM, Snyder AN, Miocinovic S. Deep brain stimulation creates an informational lesion of the stimulated nucleus. Neuroreport 2004; 15:1137–1140

15. Hashimoto T, Elder CM, Okun MS, Patrick SK, Vitek JL. Stimulation of the subthalamic nucleus changes the firing pattern of pallidal neurons. J Neurosci 2003; 23:1916–1923

16. Hodaie M, Wennberg RA, Dostrovsky JO, Lozano AM. Chronic anterior thalamus stimulation for intractable epilepsy. Epilepsia 2002; 43: 603–608

17. Mayberg HS, Lozano AM, Voon V, McNeely HE, Seminowicz D, Hamani C, Schwalb JM, Kennedy SH. Deep brain stimulation for treatment-resistant depression. Neuron 2005; 45: 651–660

18. Meissner W, Leblois A, Hansel D, Bioulac B, Gross CE, Benazzouz A, Boraud T. Subthalamic high frequency stimulation resets subthalamic firing and reduces abnormal oscillations. Brain 2005; 128: 2372–2382

19. McIntyre CC, Mori S, Sherman DL, Thakor NV, Vitek JL. Electric field and stimulating influence generated by deep brain stimulation of the subthalamic nucleus. Clin Neurophysiol 2004; 115: 589–595

20. Nuttin B, Cosyns P, Demeulemeester H, Gybels J, Meyerson B. Electrical stimulation in anterior limbs of internal capsules in patients with obsessive-compulsive disorder. Lancet 1999; 354: 1526

21. Visser-Vandewalle V, Temel Y, Colle H, van der Linden C. Bilateral high-frequency stimulation of the subthalamic nucleus in patients with multiple system atrophy—parkinsonism. Report of four cases. J Neurosurg 2003; 98: 882–887

22. Welter ML, Houeto JL, Bonnet AM, Bejjani PB, Mesnage V, Dormont D, Navarro S, Cornu P, Agid Y, Pidoux B. Effects of high-frequency stimulation on subthalamic neuronal activity in parkinsonian patients. Arch Neurol 2004; 61: 89–96

Cortical Stimulation for Movement Disorders

S. Palfi

A novel concept in movement disorders has emerged postulating that the cortex may be implicated in the genesis of motor symptoms. Further experimental and clinical studies in Parkinson's disease showed that motor cortex may also pattern abnormal rhythmic activity in the basal ganglia that underlies the observed motor symptoms. Here, we describe the use of electrical interference of motor cortex in a primate model of Parkinson's disease. Using high-frequency trains of pulses, motor cortex stimulation significantly reduces akinesia and bradykinesia in MPTP baboons. This behavioral benefit was associated with an increase metabolic activity in the supplementary motor area using 18-F-deoxyglucose Pet-scan, normalization of mean firing rate in internal globus pallidus (GPi) and subthalamic nucleus (STN) and reduction of synchronized oscillatory neuronal activities in the STN and GPi using unitary neuronal recordings. All these functional effects were similar to those reported for deep brain stimulation in STN or GPi, using intraparenchymal electrodes. Most importantly, the present series of experiments were conducted as a pre-clinical study assessing the behavioral benefit in a chronic MPTP primate model in which dopamine depletion was progressive and regularly documented using 18-F-dopa Pet-scan. The data also suggest that motor cortex stimulation was more effective in severely disabled animals implying therapeutic potential for advanced parkinsonian patients. Moreover, the major advantage of such a surgical approach is the simplicity and safety of the procedure entailing an epidural electrode that can be introduced without a deep brain stereotaxic surgery.

Placement of Subthalamic DBS Electrodes in a Radiology Suite Using Interventional MRI

P.A. Starr, A.J. Martin, P. Talke, J. Ostrem, W.K. Sootsman, N. Levesque, J. Meyers, P.S. Larson

Introduction

Standard methodology for placement of deep brain stimulator (DBS) electrodes includes preoperative stereotactic brain imaging, supplemented by intra-operative physiology in awake patients. However, multiple brain penetrations may be necessary in order to achieve an acceptable micro-electrode signature, and not all patients can tolerate awake surgery. In this study, we investigate the use of direct real-time image guidance as a means of delivering DBS electrodes. The technique may be performed on anesthetized patients and does not require stereotaxy, microelectrode recordings, or an operating theater. The devices and methods used in this study evolved from earlier work on interventional MRI guided brain biopsy [1, 2].

Methods

Imaging was performed on a 1.5 T magnet (Philips Intera, Best, The Netherlands) in a radiology suite equipped with an in-room console. All patients signed an informed consent form that was approved by the university's committee on human research. Following induction of general anesthesia, the patient's head was fixed to the MR table-top using an MRI-compatible headholder (Malcolm-Rand). A radiofrequency coil consisting of two 20 cm diameter circular loops was placed bilaterally against the patients head. A burr-hole was created and a trajectory guide (Nexframe, Image Guided Neurologics, Melbourne, FL) was attached to the skull. The trajectory guide is identical to that used for "frameless" neuronavigation-assisted DBS placement [3]. A fluid filled 14 cm long "stem" was inserted into the trajectory guide to indicate its orientation.

Patients were then transferred to magnet iso-center and high resolution T2-weighted spin echo images were acquired to delineate the STN and establish the pivot point of the trajectory guide. The path between the selected target in the dorsolateral STN and the pivot point of the trajectory guide represented the desired insertion path for the DBS lead. Trajectory guide alignment was achieved with a fluoroscopic acquisition in a scan plane perpendicular to and centered on the desired trajectory. This scan plane was positioned approximately 10 cm from the patient's skull, where only the tip of the fluid filled stem would be visible. The neurosurgeon then reached into the magnet and adjusted the trajectory guide until the stem was centered in this imaging plane (◘ Fig 9.1a). The trajectory guide was then locked and two orthogonal acquisitions were performed along the desired trajectory to assure correct orientation of the guide.

Following alignment of the trajectory guide, the fluid filled stem was removed and a peel-away introducer sheath with a titanium stylet was advanced to the STN. Confirmation scans along the desired tra-

9

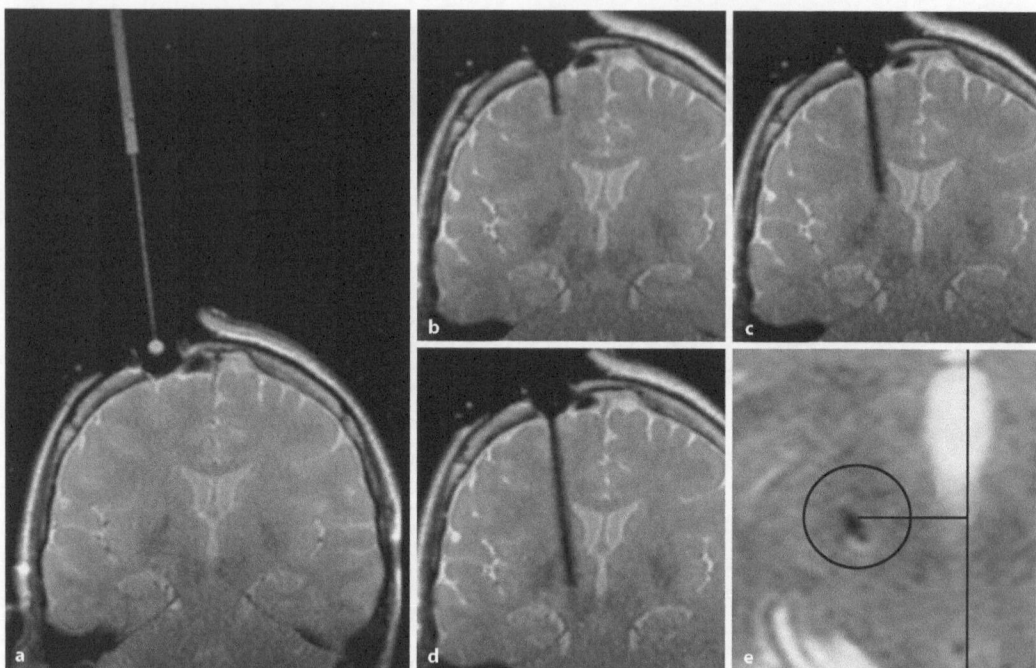

⬛ Fig. 9.1a–e. The orientation of the trajectory guide is demonstrated following alignment (**a**). Insertion of a rigid titanium stylet is monitored to confirm its trajectory and screen for hemorrhage (**b–d**). Following placement, stylet position is compared with the intended target, which is indicated by the center of the red circle (**e**)

jectory were performed during this introduction process (⬛ Fig. 9.1b–d). After confirmation of successful placement in the dorsolateral STN (⬛ Fig. 9.1e), the stylet was removed and a 28 cm long DBS lead (DBS lead model 3389, Medtronic, Minneapolis, MN) was advanced to the target through the sheath. The high resolution T2-weighted spin echo sequence was then repeated to confirm lead placement and rule out hemorrhage. Finally, the peel away sheath was removed and the lead was anchored to the skull (Stim-lock, Image Guided Neurologics). The lead extender and pulse generator were placed one week later in a standard operating room.

The "radial error" was measured on the MR console as the distance between the selected target and the actual position of the lead, in the axial plane. Clinical outcomes were assessed using the Unified Parkinson's Disease Rating Scale (UPDRS) part III, preoperatively and 6 months after DBS.

Results

Twelve leads were implanted in eight patients. Four patients had unilateral implantation, four had bilateral implantation performed as staged (n=3) or simultaneous (n=1) procedures. The mean (± SD) age of the patients was 57 ± 13 years, and the mean baseline UPDRS-III score was 20 ± 6 on medication and 58 ± 15 off medication. Eleven of twelve leads were placed using a single brain penetration; in one case a second brain penetration was required as the initial placement of the stylet/sheath was 2.5 mm from the intended target. The total operative time for the procedure decreased from 5 hours, 40 minutes for the first unilateral lead implant to 2 hours, 45 minutes for the most recent unilateral implant. The mean (± SD) radial error for lead placement was 1.2 ± 0.6 mm, range 0.1 to 2.1 mm. At 6 months, improvement from baseline in the off medication UPDRS-III score for the first two bilaterally implanted patients was 81% and 68%. Two of the eleven procedures were followed by scalp wound infections requiring lead removal.

Conclusions

Early experience suggests that placement of STN DBS electrodes in a radiology suite, using high field iMRI without physiology is accurate. Improvement in parkinsonian motor signs at 6 months following implantation is comparable to the results of STN-DBS using standard stereotaxy with microelectrode guidance. Infection is a serious concern in this environment. To address the infection risk, we have modified our procedures for initial patient draping and exposure. A more complete description of this procedure has been published [4].

References

1. Hall WA, Liu H, Martin AJ, Maxwell RE, Truwit CL. Brain biopsy sampling by using prospective stereotaxis and a trajectory guide. J Neurosurg 2001; 94: 67–71
2. Hall WA, Martin AJ, Liu H, Nussbaum ES, Maxwell RE, Truwit CL. Brain biopsy using high-field strength interventional magnetic resonance imaging. Neurosurgery 1999; 44: 807–814
3. Holloway K, Gaede S, Starr PA, Rosenow J, Ramakrishnan V, Henderson J. Frameless stereotaxy using bone fiducials for deep brain stimulation. J Neurosurg 2005 (in press)
4. Martin A, Larson P, Ostrem J, Sootsman K, Weber O, Lindsey N, Myers J, Starr PA. Placement of deep brain stimulator electrodes using real-time high field interventional MRI. Magn Res Med 2005; 54: 1107–1114

Part III:

Clinical Research in Interventional Neuro-science

Treatment of Chronic Neuropathic Pain by Motor Cortex Stimulation – Results of a Prospective Controlled Trial

J.-P. Nguyen, F. Velasco, J.-P. Lefaucheur, M. Velasco, P. Brugières, B. Boleaga, F. Brito, Y. Kéravel

Introduction

Chronic motor cortex stimulation (MCS) is a therapeutic procedure that is increasingly used for neuropathic pain refractory to medical treatment. However, no prospective controlled trial of the procedure has been published. Its real impact on the patient's improvement therefore remains controversial.

Method

This study was conducted at two different centers: the Henri Mondor Hospital, Department of Neurosurgery, in Créteil (France) and the Mexico General Hospital, Department of Stereotaxis and Functional Neurosurgery (Mexico). Each center recruited five patients with neuropathic pain refractory to medical treatment, and the final group was composed of six men and four women aged 29–75 years (mean: 54.7 ±18 years). Three patients suffered from central pain (1, 3, 8), three from neuropathic facial pain (2, 4, 5), two from post-herpetic peripheral pain (6 and 7), and two from complex regional pain syndrome (CRPS) of the upper limb (9 and 10) (◘ Table 10.1).

The patients were evaluated preoperatively and postoperatively (D-30, D30, D60, D75, D90, M6, M9, M12) using a 5-point verbal scale, a visual analogue scale, the McGill Pain Questionnaire, the Wisconsin Brief Pain Questionnaire, and the Medication Quantification Scale. At the end of the second postoperative month, patients were randomized to two groups. In the first group, the neurostimulator was turned off between D60 and D75 (OFF period) and then turned on between D75 and D90 (ON period); the opposite sequence was applied in the second group.

Results

All scores were significantly improved by the first month and remained significantly improved until the final evaluation at M12. In addition to the total score, the various components of the MPQ were also analyzed. Scores 2 and 3, evaluating affective and cognitive aspects of pain, were the most significantly improved ($p<0.02$ and $p<0.01$). Activities of daily life evaluated by certain items of the MPQ (ADL item) and WBPQ (activity item) were also significantly improved ($p<0.02$).

Overall, patients 5 and 6 showed the least marked improvement for all scores (mean variation of all scores between D-30 and M12 of –1.9% and 15.6%). Patient 3 presented very variable improvements for the various scores (mean of 39.7%). In all other patients (1, 2, 4, 7, 8, 9, and 10), practically all scores evolved in the same direction (mean improvements of 35.1%, 68.1%, 86.4%, 69.4%, 82%, 67.7%, and 69.9%,

◻ Table 10.1. Patients' clinical data

Case	Sex	Age	Etiology of pain	Distribution of pain	History [years]	Sensibility disturbance
1	M	70	Stroke (hemorrhagic)	Right hemibody	5	Hyperesthesia
2	F	75	Trigeminal neuropathy	Right hemiface	5	Allodynia
3	M	57	Stroke (ischemic)	Right hemibody	5	Hypoesthesia
4	M	57	Trigeminal neuropathy	Left hemiface	6	Hypoesthesia
5	F	31	Trigeminal neuropathy	Right hemiface	3	Hypoesthesia
6	F	75	Peripheral (herpes)	Left intercostal T5-T6	4	Allodynia
7	M	68	Peripheral (herpes)	Right C2-C3	4	Hyperesthesia
8	M	52	Stroke (ischemic)	Left hemiface	1	Allodynia
9	M	29	Peripheral (CRPS)	Left C3-T2	14	Allodynia
10	F	33	Peripheral (CRPS)	Left C5-C8	6	Allodynia

respectively). Six patients can therefore be considered to be markedly improved (2, 4, 7, 8, 9, and 10), two patients to be moderately improved (1 and 3), and two not improved (5 and 6).

The scores did not return to preoperative values during the OFF period, suggesting a long period of post-effect. Nevertheless, a significant improvement of the scores was observed during the ON period (versus OFF period). During this blind evaluation period, the condition of all patients deteriorated slightly during stimulation, although the difference was not significant compared with the score recorded at the second month, except for the MPQ score and slightly for the VAS score. After the randomization period, patients took several months to regain their initial improvement and some patients never did. It is diffi-cult to analyze the influence of the sequence order (ON then OFF versus OFF then ON) in this short series. Overall, scores during the OFF period reflected a deterioration compared with scores recorded during the ON period, except for the VAS in the ON then OFF group and the WBPQ in the OFF then ON group. No complication related to the technique was observed.

Conclusion

The results of this study strongly suggest that motor cortex stimulation markedly contributes to the improvement of patients in terms of pain intensity, activities of daily life, and analgesic consumption.

Neuromodulatory Approaches to the Treatment of Trigeminal Autonomic Cephalalgias (TACs)

P.J. Goadsby

Introduction

Trigeminal Autonomic Cephalalgias (TACs) is a grouping of headache syndromes including cluster headache, paroxysmal hemicrania and short-lasting unilateral neuralgiform headache attacks with conjunctival injection and tearing (SUNCT) [1]. These syndromes share two major clinical pictures: trigeminal distribution of pain and ipsilateral cranial autonomic symptoms [2]. These features are consistent with cranial parasympathetic activation and sympathetic hypofunction (ptosis and miosis) representing a neurapraxic effect of carotid swelling with cranial parasympathetic activation [3, 4]. The distinction between TACs and other headache syndromes is the degree of cranial autonomic activation, not its presence alone, since other primary or secondary headaches show this type of activation [5–8].

Positron emission tomography (PET) studies in cluster headache and paroxysmal hemicrania and functional magnetic resonance imaging (MRI) studies in SUNCT have demonstrated ipsilateral posterior hypothalamic activation which seems to be specific to these syndromes and not present in other types of migraine. There are direct hypothalamic-trigeminal connections [9] and the hypothalamus is known to have a modulatory role on the nociceptive and autonomic pathways, specifically trigeminovascular nociceptive pathways [10].

Summary of Background Data and State of Today's Knowledge

Cluster Headache (CH)

A CH attack is an individual episode of pain that can last from a few minutes to some hours. A cluster bout or period usually lasts from some weeks to months. Pain is excruciatingly severe and is mainly located around the orbital and temporal regions and usually lasts 45 to 90 min, with an abrupt onset and cessation. The distinctive feature of CH is the association with autonomic symptoms such as conjunctival injection, lacrimation, miosis, ptosis, eyelid edema, rhinorrhoea, nasal blockage, forehead or facial sweating. In contrast to migraine, CH sufferers are usually restless and irritable, move about and look for a movement or posture that may relieve the pain. Alcohol, nitroglycerin, exercise and elevated environmental temperature are precipitants of acute attacks. Most patients have 1 or 2 annual cluster periods, each lasting between 1 and 3 months. It is a lifelong disorder in the majority of patients.

Pharmacological Treatment. Abortive agents are administered parenterally or by the nasal route. They are represented by sumatriptan, zolmitriptan, oxygen inhalation, intranasal dihydroergotamine and subcutaneous octreotide [11–16]. For prevention of attacks, the main stay treatments are high-dose verapamil [17, 18], lithium [19], methysergide [20], melatonin [21], prednisolone [22] and topiramate [23]. Nerve blocks by means of the local injection of anesthetics and corticoids around the greater occipital nerve has yielded inconsistent results [24–26].

Surgery. Destructive procedures may be employed as a last-resort measure in patients resistant to pharmacological treatment and exclusively with unilateral headache. The main procedures employed are the interruption of the trigeminal sensory or autonomic pathways, trigeminal sensory rhizotomy via a posterior fossa approach, radiofrequency trigeminal gangliorhizolysis and microvascular decompression of the trigeminal nerve with or without microvascular decompression of the nervus intermedius. All these techniques bring with them several complications.

Neuromodulation. Franzini et al. [27] and Leone et al. [28] reported favorable lasting results in a cohort of patients with chronic CH treated with deep brain stimulation (DBS). Suboccipital nerve stimulation is also being investigated in CH [28, 29].

SUNCT

SUNCT syndrome manifests as a unilateral headache which occurs in association with cranial autonomic features. In contrast to other TACs, attacks have a very brief duration, can occur very frequently and the prominent feature is represented by conjunctival injection and lacrimation [30, 31]. This syndrome is rare, has a male predominance (sex ratio: 2.1:1) and the onset is between 40 and 70 years. Pain is usually maximal in the ophthalmic distribution of the trigeminal nerve (orbital and peri-orbital regions, forehead and temple) and is typically unilateral. It is of moderate to severe intensity and attacks last between 5 and 250 seconds [32] although attacks lasting up to 2 hours have been described [33–35]. Attacks are always accompanied by ipsilateral conjunctival injection and lacrimation, while other autonomic symptoms are less commonly reported.

The majority of patients can precipitate attacks by touching trigger zones innervated by trigeminal nerve. SUNCT must be differentiated from trigeminal neuralgia, primary (idiopathic) stabbing headache and paroxysmal hemicrania.

Pharmacological Treatment. Several categories of drugs used in other headache syndromes such as non-steroidal anti-inflammatory drugs (including indomethacin), paracetamol, 5-HT agonists (triptans, ergotamine and dihydroergotamine) beta-adrenergic blockers, tricyclic antidepressants, Ca-channel blockers (verapamil, nifedipine), methysergide, lithium, prednisolone, phenytoin, baclofen and i.v. lignocaine have proved to be ineffective in SUNCT [34]. Intravenous lignocaine has been shown to be effective in the acute suppression of SUNCT [37].

Prevention of attacks may be obtained with carbamazepine [33, 34, 36, 38, 39], lamotrigine [40–44], gabapentin [45–47] and topiramate [34].

Surgery. Several surgical approaches have been tried in SUNCT syndrome, but none has a follow up sufficient to reach reliable conclusions. Destructive procedures are therefore at the moment not recommended.

Neuromodulation. Leone et al. [48] have recently reported excellent results in SUNCT after stimulation of posterior hypothalamus.

Future Directions

Neuromodulation appears to be a promising new approach to treatment of both CH and SUNCT syndromes refractory to other treatments.

References

1. Headache Classification Committee of The International Headache Society. The International Classification of Headache Disorders (2nd edn). Cephalalgia 2004; 24 (Suppl 1):1–160
2. Goadsby PJ, Lipton RB. A review of paroxysmal hemicranias, SUNCT syndrome and other short-lasting headaches with autonomic features, including new cases. Brain 1997; 120: 193–209
3. Ekbom K, Greitz T. Carotid angiography in cluster headache. Acta Radiologica 1970; 10: 177–186
4. May A, Buchel C, Bahra A, Goadsby PJ, Frackowiak RSJ. Intracranial vessels in trigeminal transmitted pain: a PET Study. Neuroimage 1999; 9: 453–460
5. Barbanti P, Fabbrini G, Pesare M, Vanacore N, Cerbo R. Unilateral cranial autonomic symptoms in migraine. Cephalalgia 2002; 22: 256–259
6. Benoliel R, Sharav Y. Trigeminal neuralgia with lacrimation or SUNCT syndrome? Cephalalgia 1998; 18: 85–90
7. Goadsby PJ, Edvinsson L, Ekman R. Cutaneous stimulation leading to facial flushing and release of calcitonin gene-related peptide. Cephalalgia 1992; 12:53–56
8. Goadsby PJ, Matharu MS, Boes CJ. SUNCT syndrome or trigeminal neuralgia with lacrimation. Cephalalgia 2001; 21: 82–83

9. Malick A, Burstein R. Cells of origin of the trigeminohypothalamic tract in the rat. J Comp Neurol 1998; 400: 125–144
10. Bartsch T, Levy MJ, Knight YE, Goadsby PJ. Differential modulation of nociceptive dural input to [hypocretin] Orexin A and B receptor activation in the posterior hypothalamic area. Pain 2004; 109:367–378
11. Ekbom K, The Sumatriptan Cluster Headache Study Group. Treatment of acute cluster headache with sumatriptan. N Engl J Med 1991; 325:322–326
12. Cittadini E, May A, Straube A et al. Zolmitriptan nasal spray is effective and well tolerated in the acute treatment of cluster headache: a double-blind placebo-controlled crossover study. Europ J Neurology 2005 (in press)
13. Fogan L. Treatment of cluster headache: a double blind comparison of oxygen vs air inhalation. Arch Neurology 1985; 42: 362–363
14. Kitrelle JP, Grouse DS, Seybold ME. Cluster headache: local anesthetic abortive agents. Arch Neurology 1985; 42: 496–498
15. Andersson PG, Jespersen LT. Dihydroergotamine nasal spray in the treatment of attacks of cluster headache. Cephalalgia 1986; 6: 51–54
16. Matharu MS, Levy MJ, Meeran K, Goadsby PJ. Subcutaneous octreotide in cluster headache-randomized placebo-controlled double-blind cross-over study. Ann Neurology 2004; 56: 488–494
17. Lance JW, Goadsby PJ. Mechanism and Management of Headache, 7th edn. New York: Elsevier, 2005
18. Krabbe A, Steiner TJ. Prophylactic treatment of cluster headache. In: Sjaastad O, Nappi G, eds. Cluster headache syndrome in general practice: Basic concepts. London: Smith-Gordon, 2000, pp 91–96
19. Ekbom K, Solomon S. Management of cluster headache. In: Olesen J, Tfelt-Hansen P, Welch KMA (eds) The Headaches, 2nd ed. Philadelphia: Lippincott, Williams & Wilkins, 2000, pp 731–740
20. Curran DA, Hinterberger H, Lance JW. Methysergide. Res Clin Stud Headache 1967; 1:74–122
21. Leone M, D'Amico D, Moschiano F, Fraschini F, Bussone G. Melatonin versus placebo in the prophylaxis of cluster headache: a double-blind pilot study with parallel groups. Cephalalgia 1996; 16: 494–496
22. Jammes JL. The treatment of cluster headaches with prednisone. Dis Nerv Sys 1975; 36: 375–376
23. Rapoport AM, Bigal ME, Tepper SJ, Sheftell FD. Treatment of cluster headache with topiramate: effects and side-effects in five patients. Cephalalgia 2003; 23: 69–70
24. Anthony M. Arrest of attacks of cluster headache by local steroid injection of the occipital nerve. In: Rose FC (ed) Migraine: Clinical and Research Advances. London: Karger, 1985, 169–173
25. Ambrosini A, Vandenheede M, Rossi P, Aloj F, Sauli E, Buzzi MG, et al. Suboccipital (GON) injection with long-acting steroids in cluster headache: a double-blind placebo-controlled study. Cephalalgia 2003; 23: 734
26. Peres MFP, Stiles MA, Siow HC, Rozen TD, Young WB, Silberstein SD. Greater occipital nerve blockade for cluster headache. Cephalalgia 2002; 22: 520–522
27. Franzini A, Ferroli P, Leone M, Broggi G. Stimulation of the posterior hypothalamus for treatment of chronic intractable cluster headaches. The first reported series. Neurosurgery 2003; 52: 1095-1101
28. Leone M, Franzini A, Broggi G, Dodick D, Rapoport A, Goadsby PJ, et al. Deep brain stimulation for intractable chronic cluster headache: proposals for patient selection. Cephalalgia 2004; 24: 934–937
29. Weiner RL, Reed KL. Peripheral neurostimulation for control of intractable occipital neuralgia. Neuromodulation 1999; 2: 217–222
30. Sjaastad O, Russell D, Horven I, Bunnaes U. Multiple neuralgiform unilateral headache attacks associated with conjunctival injection and appearing in clusters. A nosological problem. Proceedings of the Scandinavian Migraine Society. Arhus, 1978, 31
31. Sjaastad O, Saunte C, Salvesen R, Fredriksen TA, Seim A, Roe OD, et al. Shortlasting unilateral neuralgiform headache attacks with conjunctival injection, tearing, sweating, and rhinorrhea. Cephalalgia 1989; 9: 147–156
32. Dodick DW, Trentman T, Zimmerman R, Eross EJ. Ocipital nerve stimulation for intractable chronic primary headache disorders. Cephalalgia 2003; 23: 701
33. Raimondi E, Gardella L. SUNCT syndrome. Two cases in Argentina. Headache 1998; 38: 369–371
34. Matharu MS, Boes CJ, Goadsby PJ. SUNCT syndrome: prolonged attacks, refractoriness and response to topiramate. Neurology 2002; 58: 1307
35. Pareja JA, Joubert J, Sjaastad O. SUNCT syndrome. Atypical temporal patterns. Headache 1996; 36: 108–110
36. Pareja JA, Kruszewski P, Sjaastad O. SUNCT syndrome: trials of drugs and anesthetic blockades. Headache 1995; 35: 138–142
37. Matharu MS, Cohen AS, Goadsby PJ. SUNCT syndrome responsive to intravenous lidocaine. Cephalalgia 2004; 24: 985–992
38. Ertsey C, Bozsik G, Afra J, Jelencsik I. A case of SUNCT syndrome with neurovascular compression. Cephalalgia 2000; 20: 325
39. Peatfield R, Bahra A, Goadsby PJ. Trigeminal-autonomic cephalgias (TACs). Cephalalgia 1998; 18: 358-361
40. Leone M, Rigamonti A, Usai S, D'Amico D, Grazzi L, Bussone G. Two new SUNCT cases responsive to lamotrigine. Cephalalgia 2000; 20: 845–847
41. D'Andrea G, Granella F, Ghiotto N, Nappi G. Lamotrigine in the treatment of SUNCT syndrome. Neurology 2001; 57: 1723–1725
42. Gutierrez-Garcia JM. SUNCT syndrome responsive to lamotrigine. Headache 2002; 42: 823–825
43 Chakravarty A, Mukherjee A. SUNCT syndrome responsive to lamotrigine: documentation of the first Indian case. Cephalalgia 2003; 23: 474–475
44 Piovesan EJ, Siow C, Kowacs PA, Werneck LC. Influence of lamotrigine over the SUNCT syndrome: one patient follow-up for two years. Arq Neuropsiquiatr 2003; 61: 691–694
45 Graff-Radford SB. SUNCT syndrome responsive to gabapentin. Cephalalgia 2000; 20: 515–517
46 Porta-Etessam J, Martinez-Salio A, Berbel A, Benito-Leon J. Gabapentin (neurontin) in the treatment of SUNCT syndrome. Cephalalgia 2002; 22: 249
47 Hunt CH, Dodick DW, Bosch P. SUNCT responsive to gabapentin. Headache 2002; 42: 525–526
48 Leone M, Franzini A, D'Andrea G, Broggi G, Casucci G, Bussone G. Deep brain stimulation to relieve drug-resistant SUNCT. Ann Neurol 2005; 57: 924–927

Hypothalamic Deep Brain Stimulation for the Treatment of Chronic Cluster Headaches: A Series Report

A. Franzini, P. Ferroli, M. Leone, G. Bussone, G. Broggi

Introduction

Chronic cluster headache (CH) is one of the most severe facial pain syndromes. Pain usually starts in or around the eye or the temple and may occasionally affect also the face, the neck, the eye or the entire hemicranium. Attacks are generally unilateral and start with sudden, deep, non-fluctuating and excruciating pain, which shifts to the contra-lateral side in about 15% of patients. Attacks may last from 15 min to 3 h, range in frequency from 8/day to 1/week, and occur 5 to 10 times daily in severe chronic forms. Pain is accompanied by autonomic symptoms such as lacrimation from the eye in the affected side, nasal discharge, eye reddening and sweating. Pain attacks may usually be triggered by the sublingual administration of 1 mg nitroglycerin.

CH has traditionally been considered as a vascular headache but there is clinical evidence suggesting that vascular reactions observed during the attacks are primarily due to CNS discharge. Posterior hypothalamus has been recently identified as the possible central generator of pain: positron emission tomography (PET) has shown activation in the ipsilateral inferior hypothalamic gray matter during CH attacks [1] and morphometric magnetic resonance imaging (MRI) has demonstrated an increase in neuronal hypothalamic size and density in CH patients [2]. In 1970, Sano [3] performed posterior ipsilateral hypothalamotomy to treat cancer facial pain. Since the target area was close to the hypothalamic area evidenced by PET during CH, this prompted the authors to investigate the effects of hypothalamic stimulation in CH patients and stereotactic hypothalamic surgery was successfully performed in the first patient in July 2000 [4].

Methods

Since July 2000, seven additional patients have been undergone to stereotactic hypothalamic surgery. The patient selection was performed by a co-operative team of neurologists experienced in headache and of neurosurgeons. The initial diagnosis of CH was made according to the classification of the International Headache Society [5]. Patients were receiving treatment with a number of drugs, alone or in combination such as corticosteroids, lithium, methysergide, ergotamine, calcium channel blockers, beta-adrenergic blockers, tricyclic antidepressants, valproate, topiramate, gabapentin, melatonin and non steroidal anti-inflammatory agents. Transnasal endoscopic block of sphenopalatin ganglion was performed twice in all patients before taking into consideration more invasive surgical procedures. Patients who after at least 1 year did not have pain remission were considered to be candidates for hypothalamic surgery, were examined to exclude psychiatric complications and informed on the classical surgical procedures available for treatment of intractable CH.

Of the 8 patients who underwent hypothalamic surgery since July 2000, 5 were males and 3 females, their ages ranged from 27 to 63 years (median age: 42 years), the duration of CH ranged from 1 to 7 years and the number of daily bouts ranged from 1–4 to 6–8.

Stereotactic implants were performed under local anesthesia and a pre-operative MRI was used to obtain high definition anatomic images which made possible the precise determination of both anterior commissure and posterior commissure line and of position and limits of basal ganglia and main mesencephalic nuclei. A rigid cannula was inserted through a frontal paramedian burr hole and positioned up to 10 mm from the target. This cannula was used both as a guide for micro-recording (Lead Point, Medtronic) and for placement of the definitive electrode (DBS-3389, Medtronic). After macro-stimulation to evaluate potential side effects, the guiding cannula was removed and the electrode secured to the skull, then an extension cable was connected to the electrode, tunneled subcutaneously and brought out of the skin through a stab wound. After 7–10 days of trial stimulation, the electrodes were connected to a permanent, implanted neuropulse generator (Itrel II, Medtronic) positioned subcutaneously in the subclavicular area and chronic stimulation was started after daily CH attacks reappeared. The stimulation parameters employed were: amplitude: 0.5–3.8 V, frequency: 185 Hz, pulse width: 60 microsec. Voltage was gradually increased until the therapeutic effect appeared.

Results

All patients experienced complete pain relief after 1 to 10 weeks of high-frequency hypothalamic stimulation (on average: 4.4 weeks). Three of 8 patients remained pain-free without the need of any medication, while in 5 cases attacks recurred but responded satisfactorily to low doses of methysergide or verapamil, drugs which had been completely ineffective prior to the surgical intervention. No unwanted effects attributable to chronic stimulation nor acute complications from the implant procedure were observed. There was no clinical evidence of autonomic effects related to hypothalamic stimulation: 24-hour monitoring of arterial BP showed asymptomatic orthostatic hypotension in 4 patients. In 2 cases the stimulation had to be turned off and this resulted in the sudden reappearance of CH attacks which immediately disappeared on resumption of stimulation.

The first patient who was operated in July 2000, required a contralateral second implant because of bilateral CH. Chronic stimulation of left posterior hypothalamus was successful in producing complete ipsilateral pain relief, but 8 months after right radiofrequency trigeminal rhizotomy, which had been effective in obtaining cessation of attacks, right-sided, drug refractory pain attacks recurred at a rate of 3–8 per day. After stereotactic implant in the right posterior hypothalamus and the immediate start of continuous stimulation, right sided attacks disappeared.

In another patient, who showed only a 20% decrease in attacks after the intervention, MRI showed that the electrode was 4 mm posterior to the optimal estimated target and a replacement procedure was therefore performed. A marked reduction in pain attacks occurred only a few days after the intervention.

Conclusions and Implications for Clinical Practice Today

Although a broad range of pharmacological agents is employed for treatment of CH, there are patients who develop chronic, unremitting CH refractory to any medical management. Surgical treatments based on the interruption of autonomic pathways and/or on partial or total trigeminal lesion show that success is inevitably accompanied by complications such as sensory deficit and subsequent dysesthesias, painful anesthesia, facial numbness, keratitis, etc. Recurrence rate also remains high after complete trigeminal deafferentation.

The recent PET findings during attacks and the hypothalamic abnormalities found in these patients suggest that a central dysfunction involving hypothalamic circuitry is involved in CH. Data supporting this hypothesis are:

- the effect of stimulation is strictly ipsilateral, as shown by one of our cases;
- the correct placement of the electrode in the posterior hypothalamus is mandatory to obtain a satisfactory result;
- in CH patients, opiates are not effective, ruling out a generic analgesic effect due to the release of endogenous opiates;
- the prolonged duration of pain relief in the absence of the development of tolerance, which on the contrary appears in patients undergoing periacqueductal gray matter stimulation for different types of pain [6].

Contrary to the findings of Sano [3] no undesirable autonomic responses were observed in these patients.

Future Directions

The cases reported in this paper represent the largest series of patients with chronic CH, successfully treated with high-frequency hypothalamic stimulation.

The results described suggest that this technique may represent an effective and safe treatment of CH and that it can also be employed bilaterally in case of attacks affecting both sides of the cranium. Dedicated software is being developed at Istituto Neurologico Besta to facilitate a precise placement of the electrode during the stereotactic intervention for target planning.

References

1. May A, Bahra A, Buchel C, Frackoviak RS, Goadsby PJ. Hypothalamic activation in cluster headache attacks. Lancet 1998; 352: 275–278
2. May A, Bahra A, Buchel C, Frackowiak RJS, Goadsby PJ. PET and MRA findings in cluster headache and MRA in experimental pain. Neurology 2000; 55: 1328–1335
3. Sano K, Mayanagi Y, Sekino H, Ogashiwa M, Ishijima B. Results of stimulation and destruction of the posterior hypothalamus in man. J Neurosurg 1970; 33: 689–707
4. Leone M, Franzini A, Bussone G. Stereotactic stimulation of posterior hypothalamic gray matter in a patient with intractable cluster headache. N Engl J Med 2001; 345: 1428–1429
5. Headache Classification Committee of the International Headache Society. Classification and diagnostic criteria for headache disorders, cranial neuralgia and facial pain. Cephalalgia 1988; 8 (Suppl 7): 1–96
6. Broggi G, Franzini A, Giorgi C, Spreafico R. Preliminary results of specific thalamic stimulation for deafferentation pain. Acta Neurochir 1984; 33 (Suppl): 497–500

Amygdalohippocampal Deep Brain Stimulation (Ah-DBS) for Refractory Temporal Lobe Epilepsy

P. Boon, K. Vonck, V. De Herdt, J. Caemaert, D. Van Roost

Introduction

Electrical seizure onset in the amygdala and hippocampus is the key feature of the medial temporal lobe epilepsy syndrome [1]. About 10% of patients with refractory epilepsy are scheduled for invasive video-EEG monitoring to localize the ictal onset zone during presurgical evaluation [2]. Chronic deep brain stimulation (DBS) electrodes are suitable for intracranial ictal onset localization in the medial temporal lobe [3]. Using DBS electrodes, we evaluated the efficacy and safety of amygdalo-hippocampal DBS (AH-DBS), following invasive localization of the ictal on-set zone, in patients with refractory temporal lobe epilepsy. Parts of this work have been previously reported [3, 4].

Patients and Methods

Ten patients with refractory epilepsy were implanted with bilateral AH-DBS electrodes and/or subdural grids for ictal onset localization and subsequent stimulation. In patients with ictal onset in the temporal lobe, AH-DBS was initiated at the side of ictal onset during an acute stimulation period with an external pulse generator. Patients in whom a significant reduction of interictal spikes and/or seizures was shown during this period were implanted with an abdominally located pulse generator. Patients were followed-up at the epilepsy clinic every 2–4 weeks.

Results

Four patients had a left-sided focal medial temporal lobe onset. Three patients had a right-sided regional medial temporal lobe onset. One patient had a bilateral regional temporal lobe onset with predominant involvement of the left side. Two patients had a left-sided regional medial temporal lobe onset. Nine out of 10 patients had a >50% reduction of interictal spikes during the initial AH-DBS trial period. In one patient who showed very infrequent spiking, seizure frequency that had significantly decreased, was used as a criterion for implantation. One patient did not meet the chronic implantation criterion and underwent a selective amygdalo-hippocampectomy. Nine out of 10 patients were implanted with an internal generator. The mean follow-up in these patients was 16 months (range: 9–25 months). One patient has been free of complex partial seizures (CPS) for 2 years and has been tapered off 2 anti-epileptic drugs (AEDs). Another patient has become seizure-free in the past 9 months; 3/10 patients have a >50% reduction in seizure frequency; 3/4 patients have been tapered off 1 AED. 2/10 patients have a reduction of 25% of seizure frequency but have still been tapered off 2 AEDs; one of these patients had bilateral seizure onset and is currently being revised for bilateral AH-DBS. In two patients no overall change in seizure frequency occurred. None of the patients reported side effects. The resected patient has been seizure free for 12 months. An asymptomatic haemorrhage at the tip of one of the four electrodes was reported in one patient, resolving spontaneously after one week on a control MR scan.

Discussion

High levels of invasiveness and relative inefficacy are major concerns and limitations of standard treatments that provide the impetus for further developing neurostimulation as a treatment for epilepsy.

Sensible approaches for DBS in refractory epilepsy are either:

- to target crucial "pacemaker" central nervous system structures (such as the thalamus or the subthalamic nucleus), or
- to interfere with the area of ictal onset itself.

Our study aimed at evaluating the efficacy of DBS in the medial temporal lobe after the ictal onset zone had been identified in this region.

Animal studies have shown abortive effects on epileptic activity when electrical fields were applied to hippocampal slices [5]. In-vivo studies in rats showed that electrical stimuli applied following a kindling stimulus ("quenching") can delay the development of the kindling process [6, 7]. Bragin et al. found that repeated stimulation of the hippocampal perforant path in the kainate rat model significantly reduced seizures [8]. In humans, preliminary short-term AH-DBS showed promising results with significant reduction of interictal epileptiform activity and seizure frequency [9].

Half of the patients treated with AH-DBS in this study had a reduction of seizure frequency of >50% allowing tapering off one or more AEDs. None of the patients reported side effects or showed changes in bedside neurological and neuropsychological testing. Results of formal neuropsychological testing comparing pre- and post-DBS results will be published shortly.

The mechanism of action (MOA) of DBS in reducing seizures remains unclarified. Some support the hypothesis that actual stimulation is not necessary to achieve efficacy and claim that efficacy is based on the lesion provoked by the insertion of the electrode ("microthalamotomy" effect) [10]. Furthermore, prolonged seizure control in patients who underwent invasive recording with conventional electrodes have been described [11]. Blinded randomization of patients to "on" and "off" stimulation paradigms following implantation during follow-up >6 months may clarify this issue and may also simultaneously clarify the potential effect of sham stimulation due to an implanted device. DBS may also act through local inhibition induced by current applied to nuclei that are involved in propagating, sustaining or triggering of epileptic activity in a specific CNS structure ("reversible functional lesion"). Apart from this "local" inhibition, the MOA of DBS may be based on the effect on projections leaving from the area of stimulation to other central nervous structures. This may be the most likely hypothesis when crucial structures in epileptogenic networks are involved. However, considering that the medial temporal lobe structures are also potentially involved in these networks it may be that targeting the ictal focus may also affect the epileptogenic network.

Conclusion

In this open pilot trial, AH-DBS significantly reduced seizure frequency during long-term follow-up without side effects. For patients who are less suitable candidates for epilepsy surgery, AH-DBS may become a valuable alternative. Randomized and controlled studies in larger patient series are mandatory to identify the potential treatment population and optimal stimulation paradigms.

References

1. Spencer SS, Guimaraes P, Katz A, Kim J, Spencer D. Morphological patterns of seizures recorded intracranially. Epilepsia 1992; 33: 537–545
2. Boon P, Vandekerckhove T, Achten E et al. Epilepsy surgery in Belgium, the Flemish experience. Acta Neurol Belg 1996; 6–18
3. Vonck K, Boon P, Achten E, De Reuck J, Caemaert J. Long-term amygdalo-hippocampal stimulation for refractory temporal lobe epilepsy. Ann Neurol 2002; 52: 556–565
4. Boon P, Vonck K, Van Roost D, Claeys P, De Herdt V, Achten E, Gossiaux F, Caemaert J. Amygdalohippocampal deep brain stimulation (AH-DBS) for refractroy temporal lobe epilepsy. Rev Neurol 2005; 161 (Suppl 1): 19–21

5. Lian J, Bikson M, Sciortino C, Stacey WC, Durand DM. Local suppression of epileptiform activity by electrical stimulation in rat hippocampus in vitro. J Physiol 2003; 547: 427–434

6. Weiss SRB, Li XL, Rosen JB, Li H, Heynen T, Post RM. Quenching: inhibition of development and expression of amygdala kindled seizures with low frequency stimulation. Neuroreport 1995; 4: 2171–2176

7. Velisek L, Veliskova J, Stanton PK. Low frequency stimulation of the kindling focus delays basolateral amygdala kindling in immature rats. Neuroscience Letters 2002; 326: 61–63

8. Bragin A, Wilson CL, Engel J. Increased afterdischarge threshold during kindling in epileptic rats. Experimental Brain Research 2002; 144: 30–37

9. Velasco M, Velasco F, Velasco AL, Boleaga B, Jimenez F, Brito F, Marquez I. Subacute electrical stimulation of the hippocampus blocks intractable temporal lobe seizures and paroxysmal EEG activities. Epilepsia 2000; 41: 158–169

10. Hodaie M, Wennberg RA, Dostrovsky J, Lozano A. Chronic anterior thalamic stimulation for intractable epilepsy. Epilepsia 2002; 43: 603–608

11. Katariwala NM, Bakay RAE, Pennel PB, Olson LD, Henry TR, Epstein CM. Remission of intractable epilepsy following implantation of intracranial electrodes. Neurology 2001; 57: 1505–1507

Non-Pharmacological Approaches to the Treatment of Depression – Mechanisms and Future Prospects

T.E. Schlaepfer

Introduction

There is growing awareness within the field of psychiatry of an urgent need for new therapeutic options for patients with treatment refractory severe depression. The recognition that psychiatric disorders such as depression and obsessive compulsive disorder are correlated with impared function of circuits within specific brain regions, many of which are becoming well characterized, has led to the development of techniques for stimulation of these brain circuits. Novel methods of brain stimulation developed over the last decade include repetitive transcranial magnetic stimulation (rTMS), magnetic seizure therapy (MST), vagus nerve stimulation (VNS) and deep brain stimulation (DBS).

Transcranial Magnetic Stimulation

Transcranial magnetic stimulation (TMS) employs a hand-held stimulating coil applied directly to the head to deliver very strong magnetic fields to the cerebral cortex in order to induce currents which are able to depolarize neurons [3]. Unlike vagus nerve stimulation, deep brain stimulation and magnetic seizure therapy, TMS requires neither an implanted prosthesis nor general anesthesia [2]. It has been demonstrated that a range of measures of brain function are influenced by TMS, including increased early

gene expression (c-FOS) in the periventricular nucleus of the thalamus, dopamine release and changes in cortisol and prolactin levels.

Studies of TMS in depression have yielded inconclusive results. A systematic review of approximately 15 randomized, placebo-controlled clinical studies involving around 200 patients reached the conclusion that there is currently insufficient evidence to suggest that TMS is effective for this indication [9]. The authors did not rule out a beneficial effect from this intervention; however, they cited a number of methodological flaws in the evidence base, for example small sample size, concurrent use of psychotropic medication and failure to conceal the treatment group to which a patient was allocated [9].

Studies support a potential antidepressant effect from repetitive TMS (rTMS), with one study assessing rTMS to the left dorsal prefrontal cortex in seven children and adolescents with depression, reporting a response in five of the seven patients treated [13]. Significant adverse events, seizures or cognitive changes with TMS have not been reported [11, 13]. Today, rTMS presents an interesting and potentially promising technique.

Magnetic Seizure Treatment

While electroconvulsive treatment (ECT) has demonstrated unparalleled efficacy in severe depression, it is associated with cognitive adverse events [15]. Improved understanding of the mechanisms underlying ECT has led to the development of magnetic seizure treatment (MST). The first use of therapeutic magnetic seizure induction in a psychiatric patient took place at the University Hospital in Bern, Switzerland, in May 2000. MST uses TMS to induce therapeutic seizures under general anesthesia in the same setting as for ECT [5]. The electrical field induced by MST for seizure induction is more focal and limited than that induced by ECT [6]. This enhanced control allows treatment to be focused on target cortical structures considered essential to the antidepressant response, while reducing spread to medial temporal regions, which are associated with the cognitive adverse events of ECT [7].

Although MST is at an early stage of development, preliminary data suggest advantages over ECT in terms of both subjective adverse events and acute cognitive function [7]. A recent randomized, within-patient, double-masked trial compared ECT and MST in 10 patients and indicated superiority with MST in terms of time to recovery of orientation, measures of attention, retrograde amnesia, category fluency and a reduced incidence of adverse events [8]. Studies are currently underway to address the antidepressant efficacy of MST [7].

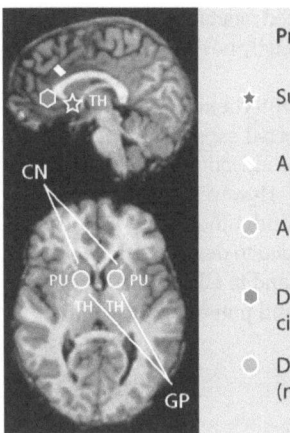

Procedures

★ Subcaudate Tractotomy

◆ Anterior cingulotomy

○ Anterior capsulotomy

● DBS of subgenual cingulate (cg25)

◐ DBS of internal capsule (nucleus accumbens)

☐ **Fig. 14.1.** Neurosurgical targets for treatment of refractory obsessive-compulsive disorder and major depression. Historically a multitargeted procedure, termed limbic leucotomy, combining anterior cingulotomy and subcaudate tractotomy has been used for intractable depression, obsessive-compulsive disorder and some other forms of severe anxiety disorders. (From Schlaepfer and Lieb [15])

resistant depression. In 59 patients, a response rate of 42% (25/59) and a remission rate of 22% (13/29) were seen at 2 years of VNS treatment in one study [12].

Vagus Nerve Stimulation

Vagus nerve stimulation (VNS) is an established treatment for drug-resistant partial-onset seizures in epilepsy and is now approved by the FDA for the treatment of refractory depression.

During VNS, electrical signals are delivered to the left vagus nerve at the cervical level. The pulse generator is implanted in a subcutaneous chest pocket just below the clavicle, whereas the electrodes are attached to the vagus nerve through an incision at the neck [4].

This treatment appears to be well tolerated by and of benefit to patients with treatment-resistant depression. In a study of 30 patients, a response rate of 40% (12/30 patients) was seen after 12 months of VNS treatment, with a remission rate of 29% (8/28 patients). This result is supported by two recent 12-month studies [2, 14]; VNS demonstrated significantly greater antidepressant benefit than usual treatment procedures at 1 year [2].

Longer-term studies also support the use of VNS in the treatment of chronic or recurrent treatment-

Deep Brain Stimulation

Deep brain stimulation (DBS) is a particularly promising investigational treatment in neuropsychiatry and is conducted through the stereotactic placement of unilateral or bilateral electrodes connected to a permanently implanted neurostimulator [15].

Although its exact mode of action is unknown, the hypothesis is that chronic high frequency (130–185 Hz) stimulation reduces neural transmission through the inactivation of voltage-dependent ion channels. Recently, promising results have been seen in refractory depression with DBS close to the subgenual cingulated region cg25 (Brodmann area 25) [10]. This area is metabolically overactive in treatment-resistant depression and DBS may reduce the elevation in activity and produce benefit in patients with treatment-resistant depression.

After 2 months of chronic white matter tract stimulation, a striking response on depression was seen in five patients, of whom four maintained this response after 6 months. Antidepressant effects were associated with a marked reduction in cerebral blood flow in cg25 as measured by positron emission tomography [10].

Other targets for DBS could also include the anterior limb of the internal capsule and the nucleus accumbens which has connections to both amygdala and prefrontal cortex. However, this treatment remains experimental and randomized, controlled, crossover data are needed to determine the optimum duration of treatment and to allay concerns over the risk of brain surgery and potentially severe adverse events with DBS [15].

Conclusions

ECT is the most effective treatment known for major depression, nevertheless it retains undeserved public stigma and is used as a last resort for non-responders to therapy. However, good evidence is emerging for promising newer methods of brain stimulation. There has been a flurry of new evidence recently, which has established VNS as a credible alternative treatment. In addition, although at an early stage, the evidence benefits from MST and DBS are compelling and await confirmation in controlled trials. Given DBS ability to very directly and focusedly modulate dysfunctional deeper brain circuits it might well be that this method will be demonstrated to be the most efficacious in the treatment of very refractory neuropsychiatric disorders.

References

1. George MS, Rush AJ, Marangell LB, Sackeim HA et al. A one-year comparison of vagus nerve stimulation with treatment as usual for treatment-resistant depression. Biol Psychiatry 2005; 58: 364–373
2. George MS, Nahas Z, Kozol FA et al. Mechanisms and the current state of transcranial magnetic stimulation. CNS Spectr 2003; 8: 496–514
3. Hallett M. Transcranial magnetic stimulation and the human brain. Nature 2000; 406: 147–150
4. Kosel M, Schlaepfer TE. Beyond the treatment of epilepsy: new applications of vagus. nerve stimulation in psychiatry. CNS Spectr 2003; 8: 515–521
5. Lisanby SH, Luber, B.; Finck, AD et al. Deliberate seizure induction with repetitive transcranial magnetic stimulation. Arch Gen Psychiatry 2001; 58: 199–200
6. Lisanby SH. Update on magnetic seizure therapy: a novel form of convulsive therapy. J ECT 2002; 18:182–188
7. Lisanby SH et al. Morales O, Payne N, et al. New developments in electroconvulsive therapy and magnetic seizure therapy. CNS Spectr 2003a; 8: 529–36
8. Lisanby SH, Luber B, Schlaepfer TE, Sackeim HA. Safety and feasibility of magnetic seizure therapy (MST) in major depression: randomized within-subject comparison with electroconvulsive therapy. Neuropsychopharmacology 2003b; 28:1852–1865
9. Martin JL, Barbanoj MJ, Schlaepfer TE, Thompson E, Perez V, Kulisevsky J. Repetitive transcranial magnetic stimulation for the treatment of depression. Systematic review and meta-analysis. Br J Psychiatry 2003; 182: 480–91
10. Mayberg HS, Lozano AM, Voon V, et al. Deep brain stimulation for treatment-resistant depression. Neuron 2005; 45: 651–60
11. Mosimann UP, Schmitt W, Greenberg BD, Kosel M, Muri RM, Berkhoff M, Hess CW, Fisch HU, Schlaepfer TE. Repetitive transcranial magnetic stimulation: a putative add-on treatment for major depression in elderly patients. Psych Res 2004; 126: 123–33
12. Nahas Z et al. Two-year outcome of vagus nerve stimulation (VNS) for treatment of major depressive episodes. J Clin Psychiatry. 2005; 66: 1097–104
13. Quintana H. Transcranial magnetic stimulation in persons younger than the age of 18. J ECT 2005; 21: 88–95
14. Rush AJ et al. Effects of 12 months of vagus nerve stimulation in treatment-resistant depression: A naturalistic study. Biol Psychiatry 2005; 58: 355–63
15. Schlaepfer TE, Lieb KL. Deep brain stimulation for treatment of refractory depression. Lancet 2005; 366: 1420–1422
16. Schlaepfer TE. Progress in therapeutic brain stimulation in neuropsychiatry. CNS Spectr 2003; 8: 488

DBS and the Treatment of Obsessive Compulsive Disorder

L. Gabriëls, B. Nuttin, P. Cosyns

Introduction

Obsessive-compulsive disorder (OCD) affects ca. 2% of the general population [1]. The cardinal symptoms of OCD are intrusive thoughts (obsessions) and/or repetitive behaviors (compulsions) that persist despite the patient's attempts to eliminate them. The obsessions and compulsions are accompanied by marked, overwhelming anxiety and are distressing and time-consuming. In addition, patients tend to avoid objects or situations that provoke obsessions or compulsions. Their functioning becomes increasingly limited by avoidance behaviors and they are burdened by shame and demoralization. Although the majority of the patients recognize the exaggerated nature of their obsessions and senselessness of their compulsions, they feel enslaved and are compelled to engage in their rituals. They cannot simply dismiss the obsessional ideas and large amounts of time are spent on compulsive rituals. The bizarre and exaggerated aspects of the symptoms result in a deep sense of shame and may lead to social isolation and depression. Co-morbidity of OCD with depression is considerable: up to 67% of patients with primary OCD have a lifetime history positive for major depressive disorder [2].

Both pharmacotherapy and cognitive behavioral psychotherapy (CBT) have proven to be effective in the treatment of OCD. However, standard treatments do not work for some patients: full or partial remission is only seen in approximately 60–80%, while the remaining patients experience only a minimal or no response. Notwithstanding the important advances made over the last decades in the efficacy, safety, and tolerance of treatments for OCD, up to 7.1% of the patients show persistent disabling symptoms in spite of combined pharmacological and psychotherapeutic treatment [3]. For these patients who remain severely disabled despite these state-of-the-art approaches new treatment strategies are needed.

Summary of Background Data and Methods

For some of these extremely suffering and treatment-refractory patients stereotactic neurosurgical intervention may be considered. Four stereotactic neurosurgical lesioning techniques have been available for the treatment of these patients: limbic leucotomy, subcaudate tractotomy, anterior cingulotomy and capsulotomy.

There is growing evidence for a neurobiological basis for OCD. Abnormalities in frontal lobe and basal ganglia function in OCD patients have led to hypotheses about the pathogenesis of the disorder [4]. One of the important loops in OCD, the frontal-striatal-pallidal-thalamic-frontal loop, passes through the anterior limb of the internal capsule, the target in anterior capsulotomy [5].

Only a limited number of procedures for neurosurgical treatment for mental disorders is carried out at a handful of centers in the world, with response rates varying between 35 and 65% [6]. While serious adverse effects are described [6–8], the overall side-effect burden is acceptable to the patient suffering previously from a severe, longstanding, intractable psychiatric disorder. Nevertheless, these side effects remain a major concern, in view of the irreversibility of the current lesioning procedures.

From scientific viewpoint, an important issue is the lack of randomized double blind "placebo-controlled" trials with these lesioning techniques. One can argue that a placebo effect in neurosurgical treatment for OCD is unlikely, bearing in mind that the psychiatric disorder has run over such a long course pre-operatively and that OCD is known to have low placebo responses to treatment. There is little evidence for spontaneous remission in severe, intractable and longstanding OCD [9]. In a follow-up of patients who are eligible for intervention but never undergo surgery for different reasons, their condition remained the same and some of them eventually committed suicide [10].

Development of deep brain stimulation (DBS) opens a new avenue for research and neurosurgical treatment in psychiatric disorders. A major advantage of stimulation, compared to conventional ablative neurosurgery, is that it is reversible. The implantation of electrodes in the brain does not significantly damage brain tissue and the stimulation itself can be modified or discontinued in the event of side effects. Electrical brain stimulation enables double blind research: once electrodes and stimulators are implanted, electrical current can be switched on or off, or can be applied at different amplitudes without the patient's knowledge. The effects of deep brain stimulation on symptoms of the psychiatric disorder, as well as on personality, cognitive and social functioning, and quality of life can be investigated in a prospective, double blind protocol.

Currently, DBS is accepted as the standard of therapy for medically refractory Parkinson's disease (PD), where it replaced the former ablative procedures. In PD, the DBS electrodes are positioned in the same area where formerly the lesioning probe was targeted. In a recent randomized prospective study clinicians confirmed the inherent advantage of DBS over their lesioning counterparts [11]. DBS had similar therapeutic benefits, fewer side effects and was superior in overall improvement of daily functioning.

We hypothesized that in the treatment of refractory OCD patients, for whom capsulotomy is considered, DBS in the anterior limbs of the internal capsules might decrease the severity of OCD. The main aim of our study is the evaluation of the clinical efficacy and safety of electrical stimulation in the anterior limbs of the internal capsules (capsular stimulation) in severe OCD.

Since DBS does not require destruction of brain tissue, it does not preclude further newer treatments should they become available. Moreover, it is possible to assess any benefits that may accrue, but if the patient does not want to continue treatment, or if severe side effects should appear that contraindicate further stimulation, it can either be switched off or even have the electrodes removed. If the therapeutic results of capsular stimulation are insufficient, removing the electrodes and performing a capsulotomy remains a therapeutic option.

Patients suffering from long-standing, severe, highly disabling OCD were screened and evaluated in the department of psychiatry of the University Hospital of Antwerpen for complete psychiatric and neuropsychological assessment, both before treatment and during the whole study period. They were referred to the University Hospitals of Leuven, Belgium, for neurosurgical intervention. Both hospitals' Ethical Review Boards approved the study protocol. Inclusion criteria required a diagnosis by Structural Clinical Interview for DSM-IV (SCID-IV) of OCD, judged to be of disabling severity, with a Yale-Brown Obsessive-Compulsive Scale (Y-BOCS) score of at least 30/40 and a Global Assessment of Functioning (GAF) score of 45 or less. This level of impairment must have persisted for a minimum of five years. Reports on failures of an exhaustive array of other available treatments for OCD are required: ineffectiveness or intolerance to adequate trials of at least 3 selective serotonin reuptake inhibitors and clomipramine, augmentation strategies with antipsychotics, and CBT. Patients had to be at least 18 old years, and no more than 60 years of age. They must be able to understand, comply with instructions and provide their own written informed consent. The patient and a close family member were repeatedly and fully informed on both procedures (capsulotomy and capsular stimulation). The standard risks known from the experience of DBS in PD were explained. Exclusion criteria were a current or past psychotic disorder, any clinically significant disorder or medical illness affecting brain function or structure (other than motor tics or Gilles de la Tourette syndrome), or current or unstably remitted substance abuse.

Medication was tapered off to an attainable minimum at least 6 weeks before surgical intervention and kept at a constant level for the first year after electrode implantation. Baseline assessment was accomplished after stabilization of the pharmacological treatment. The primary outcome measure is the Yale-Brown Obsessive-Compulsive Scale score (Y-BOCS) [12,13].

Neurosurgical intervention on all patients was performed by the same neurosurgeon (Bart Nuttin). We described this procedure elsewhere [14, 15]. Initially, the targets in the internal capsules were similar to those aimed for in the anterior capsulotomy [16]. The electrode was extended ventrally to the most inferior capsular fibers. The most distal contact was very near to or in the nucleus accumbens, the next two contacts were situated in the internal capsule, and the most proximal contact (the one farthest away from the electrode tip) was sited dorsally to the internal capsule.

After an initial stimulation phase of variable length, when stimulation parameters where screened and contact combinations that yielded therapeutic benefits were searched, patients entered a randomized long-term double blind crossover design: a branch with stimulator continuously "on" during 3 months was followed by a branch with stimulator "off" during 3 months or vice versa in random order. At the end of each branch, psychiatric assessment were completed. Patients and evaluators were blinded for the stimulation condition. However, if the assessors documented a serious worsening in the patient's condition, if suicidal ideation appeared or if patient and assessors decided that the suffering of the patient was unbearable any longer, measures were taken to shorten that branch of the crossover and to switch to the other condition, without unblinding the patient or the assessors.

Results

Eleven patients have been included in our protocol and have received bilateral electrode implants in the anterior limbs of the internal capsules.

With varying stimulation-parameters, a whole range of acute stimulation effects were observed. Some of them have been reported elsewhere [17]. Paresthesias or a warm feeling and flushing were observed in all patients although contact combinations and threshold level at which they appeared differed. Brusque abolition of stimulation frequently caused a transient hot feeling, transpiration and flushing as well.

Changes in affect were most prominent under bilateral stimulation. All patients reported sudden happiness, joy and a good feeling some seconds after stimulation was switched on with particular contact combinations. They smiled and laughed extensively.

When asked why they were laughing they often could not give a special reason. They just felt an inner joy and were filled with silent laughter. They sometimes tried to swallow their laughter but this ended in a laughing fit. Patients became more talkative and talked in a louder voice when happy feelings were induced. In at least 4 patients unilateral stimulation (both left and right) with the deepest contacts produced transient contralateral contraction of facial muscles resulting in a typical demi-smile with higher amplitudes. At the same time, they reported a jolly feeling. Bilateral stimulation of the same contacts produced less reproducible involuntary muscle contraction although it was witnessed at some occasions. In three patients some contact combinations led to a worsening mood, depressive feelings and more anxiety. Switching stimulation off or changing to other contact combinations reversed those feelings.

Other transient effects with varying stimulation-parameters include verbal perseveration, dysarthria, hyperventilation, nausea, sudden epigastric sensations, peculiar feeling in the throat, prolonged muscle contraction in cheek and neck leading to cramp. Both bilateral and unilateral stimulation with the deepest contacts produced a transient smell sensation in 2 patients. Four patients reported transient visual perceptions (black or white specks, and subtle changes in the color of the walls, impression of moving objects, or they saw everything blurred and deformed).

For eight patients double blind assessment of YBOCS scores during stimulation "on" and stimulation "off" were obtained. Mean Y-BOCS (SD) at baseline before surgery was 33.8 (3.2). At the end of the crossover branch, during which electrical stimulation was switched off, mean Y-BOCS (SD) was 33.1 (2.7), while at the end of the crossover branch during which patients received electrical stimulation mean Y-BOCS (SD) was 17.0 (7.8). In six of these eight patients the severity of OCD, as measured by the Y-BOCS, decreased of more than 35% and thus they were considered responders.

Side effects include changes in weight and sleep pattern. The changes in sleep pattern are reversible with stimulation. Other side effects (disinhibition, overconfidence, inaccurate risk assessment) are amplitude dependent and disappear when the amplitude is lowered, but this sometimes comes at the cost of the therapeutic effect on OCD and requires careful balancing between therapeutic aim and undesirable secondary effects.

Due to the high current densities necessary to obtain optimal therapeutic benefit, battery life is currently restricted. Since obsessions and compulsive rituals typically return within hours to days after failure of the batteries, patients require regular replacements. Obsessions and symptoms return with former intensity and often patients become severely depressed as well.

In an attempt to optimize the target for stimulation and thus decrease the necessary current density, the optimal area for electrical stimulation within the region of the anterior limbs of internal capsules and the grey matter caudal to this area was investigated. Therefore, the reduction in Y-BOCS when patients where stimulated compared to no stimulation was calculated and expressed as a percentage of the Y-BOCS score when stimulation was off. These reductions where highly correlated to the distance of the implanted electrode to the posterior border of the anterior commissure, both at the left (Kendall's tau B=0.889; p=0.003) and at the right side (Kendall's tau B=0.926; p=0.002), at the level of the anterior commissure, in an axial plane approximately through AC-PC. For the patients with the largest reductions (i.e. the most therapeutic benefit), the electrode was situated 1 mm posterior to the posterior border of the anterior commissure at the level of AC-PC. Correlations with the laterality of the electrode tip to the midline were not statistically significant. In the patients with optimal therapeutic benefits, the tips were located between 6 and 8 mm lateral of the midline and the stimulating contacts are situated in the Bed nucleus of the stria terminalis and the great terminal island.

Conclusions and Future Directions

Treatment-refractory OCD patients considered for neurosurgical treatment have a longstanding history of extremely persistent and incapacitating intrusive obsessions and repetitive compulsions. OCD symptoms dominate every aspect of daily activities and a meaningful way of living becomes impossible. In this present, multidisciplinary study, we demonstrated that electrical stimulation in the anterior limbs of the internal capsules induces clinically significant therapeutic benefit in patients with severe, treatment refractory OCD. Capsular stimulation not only leads to a substantial decrease in severity of OCD symptoms, it also has a beneficial impact on the patient's mood scores. Compared to capsulotomy, the observed side effects are acceptable.

Technical aspects currently limit the use of capsular stimulation as a therapeutic option. These may be overcome by the development of a battery with a longer life span, a rechargeable battery or further research on target optimizing.

Treatment of OCD patients with capsular stimulation remains investigational and is not considered as standard therapy. Therapeutic innovations should always be incorporated into a research project in order to establish their true efficacy and safety while retaining the therapeutic objectives. Such a study necessitates considerable commitment of multidisciplinary teams and patients.

The OCD-DBS collaborative group was established in an effort to prevent indiscriminate and widespread application of electrical brain stimulation before adequate long-term safety data are available, and to ensure adequate human subject protection while providing access to potential therapeutic innovation. We hereby stress the importance of the guidelines issued by this group [18].

References

1. Rasmussen SA, Eisen JL. The epidemiology and clinical features of obsessive compulsive disorder. Psychiatric Clin North Am 1992; 15: 743–758
2. Rasmussen SA, Eisen JL. Clinical and epidemiologic findings of significance to neuropharmacologic trials in OCD. Psychopharmacol Bull 1988; 24: 466–470
3. Zitterl W, Demal U, Aigner M, Lenz G, Urban C, Zapotoczky HG, et al. Naturalistic course of obsessive compulsive disorder and comorbid depression. Longitudinal results of a prospective follow-up study of 74 actively treated patients. Psychopathology 2000; 33: 75–80
4. Rauch SL. Neuroimaging and neurocircuitry models pertaining to the neurosurgical treatment of psychiatric disorders. Neurosurg Clin North Am 2003; 14: 213–223
5. Modell JG, Mountz JM, Curtis GC, Greden JF. Neurophysiologic dysfunction in basal ganglia/limbic striatal and thalamocortical circuits as a pathogenetic mechanism of obsessive-compulsive disorder. J Neuropsychiatry Clin Neurosci 1989; 1: 27–36
6. Jenike MA. Neurosurgical treatment of obsessive-compulsive disorder. Br J Psychiatry 1998; 35 (Suppl): 79-90
7. Albucher RC, Curtis GC, Pitts K. Neurosurgery for obsessive-compulsive disorder: problems with comorbidity. Am J Psychiatry 1999; 156: 495–496
8. Irle E, Exner C, Thielen K, Weniger G, Ruther E. Obsessive-compulsive disorder and ventromedial frontal lesions: clinical and neuropsychological findings. Am J Psychiatry 1998; 155: 255–263
9. Rasmussen SA, Tsuang MT. The epidemiology of obsessive compulsive disorder. J Clin Psychiatry 1984; 45: 450–457
10. Mindus P. Present-day indications for capsulotomy. Acta Neurochirurgica 1993; 58 (Suppl): 29–33

11. Schuurman PR, Bosch A, Bossuyt PMM, Bonsel GJ, Van Someren EJW, de Bie RMA, et al. A comparison of continuous thalamic stimulation and thalamotomy for suppression of severe tremor. NEJM 2000; 342: 461–468

12. Goodman WK, Price LH, Rasmussen SA, Mazure C, Delgado P, Heninger GR, et al. The Yale-Brown Obsessive Compulsive Scale. II. Validity. Arch Gen Psychiatry 1989; 46: 1012–1016

13. Goodman WK, Price LH, Rasmussen SA, Mazure C, Fleischmann RL, Hill CL, et al. The Yale-Brown Obsessive Compulsive Scale. I. Development, use, and reliability. Arch Gen Psychiatry 1989; 46: 1006–1011

14. Gabriels L, Cosyns P, Nuttin B, Demeulemeester H, Gybels J. Deep brain stimulation for treatment-refractory obsessive-compulsive disorder: psychopathological and neuropsychological outcome in three cases. Acta Psychiatrica Scand 2003; 107: 275–282

15. Nuttin BJ, Gabriels LA, Cosyns PR, Meyerson BA, Andreewitch S, Sunaert SG, et al. Long-term electrical capsular stimulation in patients with obsessive-compulsive disorder. Neurosurgery 2003; 52: 1263–1272; discussion 1272–1274

16. Litofsky NS, Chin LS, Tang G, Baker S, Giannotta SL, Apuzzo ML. The use of lobectomy in the management of severe closed-head trauma. Neurosurgery 1994; 34: 628–632; discussion 632–633

17. Nuttin BJ, Gabriels L, van Kuyck K, Cosyns P. Electrical stimulation of the anterior limbs of the internal capsules in patients with severe obsessive-compulsive disorder: anecdotal reports. Neurosurg Clin North Am 2003; 14: 267–274

18. OCD-DBS Collaborative Group. Deep brain stimulation for psychiatric disorders. Neurosurgery 2002; 51: 519

The Nucleus Accumbens: A Target for Deep-Brain Stimulation in Obsessive-Compulsive and Anxiety Disorders

V. Sturm, D. Lenartz, A. Koulousakis, H. Treuer, K. Herholz, J. C. Klein, J. Klosterkötter

Summary

We considered clinical observations in patients with obsessive-compulsive and anxiety disorders who underwent bilateral anterior capsulotomy, as well as anatomical and pathophysiological findings. Based on these considerations, we chose the shell region of the right nucleus accumbens as the target for deep-brain stimulation in a pilot series of four patients with severe obsessive-compulsive and anxiety disorders. Significant reduction in severity of symptoms has been achieved in three of the four patients treated. Clinical results, as well as a 15O-H2O-PET study performed in one patient during stimulation, speak in favor of the following hypothesis: As a central relay structure between the amygdala, basal ganglia, mesolimbic dopaminergic areas, mediodorsal thalamus, and prefrontal cortex, the nucleus accumbens seems to play a modulatory role in the flow of information from the amygdaloid complex to the latter areas. If disturbed, an imbalanced information flow from the amygdaloid complex can yield obsessive-compulsive and anxiety disorders. These can be counteracted by blocking the information flow within the shell region of the nucleus accumbens by means of deep-brain stimulation (DBS).

Introduction

Obsessive-compulsive disorder (OCD) is a chronic and disabling condition which severely impairs personal, social, and professional life. Patients with OCD suffer from recurrent obsessive thoughts and uncontrollable compulsive reactions, such as repetitive behavioral or mental acts occurring in response to an obsession. OCD occurs frequently in combination with other anxiety and depressive disorders. It is notorious for both the chronicity and the difficulty of its treatment. In severe cases of treatment-refractory OCD and anxiety disorders, neurosurgical procedures (cingulotomy, limbic leukotomy, subcaudate tractotomy, and anterior capsulotomy) may be indicated [10, 18]. The best results have been obtained with bilateral anterior capsulotomy [12, 15].

Electrical deep-brain stimulation (DBS) at high frequencies has a blocking effect on the stimulated area and mimics the effect of tissue lesioning [2, 3]. DBS is reversible and has a much lower rate of side effects than lesioning with thermocoagulation [21]. Thus, Nuttin and Cosyns [17] used bilateral DBS of the anterior limb of the internal capsule instead of lesioning for treatment of severe OCD. Significant improvement of symptoms was achieved in four patients. However, unusually high stimulation amplitudes had to be used, which resulted in high energy consumption requiring frequent exchange of the portable energy source.

Based on clinical observations as well as on anatomical and pathophysiological considerations, we used the right nucleus accumbens as the primary target for DBS in four patients with the diagnosis of severe, pharmaceutically resistant anxiety disorders and OCD.

Anatomy and Pathophysiology

During the past three decades, basal forebrain areas, especially the ventral striatum, the nucleus accumbens, and the rostral parts of the "extended amygdala" [4, 9], have attracted the growing interest of anatomists, pharmacologists, and clinicians. The dopamine theory of schizophrenia has focused on the nucleus accumbens and its role in psychiatric diseases [14, 23]. The nucleus is located immediately underneath the anterior limb of the internal capsule and covers a large area of the basal forebrain rostral to the anterior commissure.

Medially adjacent to it is the vertical part of the diagonal band of Broca, while laterally adjacent to it are the claustrum and piriform cortex. Dorsally, neighboring structures include rostral extensions of the globus pallidus and the anterior limb of the internal capsule. The nucleus accumbens extends dorsolaterally into the ventral putamen, dorsomedially into the ventral caudate, i.e., the ventral striatum sensu stricto, without a sharp demarcation.

The nucleus accumbens is divided into two principal parts: a central core and a peripheral shell. The former is associated with the extrapyramidal motor, the latter with the limbic system. While the core-shell dichotomy is well-established in rodents, in the primate, especially in man, both parts are poorly characterized. However, there is a consensus that the shell region is confined to the ventromedial margin of the nucleus (for review see Heimer [8]).

The shell region has histological and biochemical properties similar to those of the central amygdaloid nucleus, which, together with the medial amygdaloid nucleus, gives origin to the extended amygdala system (de Olmos and Heimer 1999; Alheid et al. 1998). It contains a larger proportion of relatively small cells with high concentrations of D1- and D3-receptors and a denser distribution of many neuropeptides such as VIP, CCK, enkephalins, substance P, and neurotensin, than other regions of the nucleus accumbens and the ventral striatum (Heimer 2000).

Within the nucleus accumbens, information is transmitted from shell to core. Together with the ventral striatum, the nucleus accumbens, especially the shell region, receives a strong dopaminergic input from the VTA and the dorsal tier of the substantia nigra and projects back to major parts of the dorsal and ventral tier (dorsal and and densocellular parts of the substantia nigra pars compacta) as described by Haber et al. [7].

In the human being, the nucleus accumbens receives strong afferents from the basolateral amygdala via the ventral amygdalofugal pathway, and most probably also from the central and medial amygdaloid nuclei, via the sublenticular and supracapsular parts of the extended amygdala [1, 4]. Its main efferents innervate the pallidum, striatum, mediodorsal thalamus, prefrontal, including cingulate cortex and, as mentioned above, mesolimbic dopaminergic areas.

The nucleus accumbens thus attains a central position between limbic and mesolimbic dopaminergic structures, the basal ganglia, the mediodorsal thalamus, and the prefrontal cortex.

Since dopamine is a major transmitter in the nucleus accumbens, a modulating function on amygdaloid-basal ganglia-prefrontal cortex circuitry can be assumed [5, 16, 19].

Implications for Psychiatric Surgery

In the 1960s, Leksell and Talairach introduced anterior capsulotomy as a treatment for severe OCD and anxiety disorders. Fiber tracts connecting the mediodorsal thalamus reciprocally with the prefrontal cortex were interrupted by thermocoagulation or focused stereotactic irradiation bilaterally in the anterior limb of the internal capsule [12, 15]. Significant reduction of OCD-related behavior, fear, and anxiety has been achieved in the majority of patients, but "frontal" symptoms have been observed occasionally, most probably because fibers projecting to the dorsolateral prefrontal cortex have been interrupted, in addition to fibers terminating in orbitofrontal regions.

Based on growing experience with DBS for Parkinson's disease [2, 24], Nuttin and Cosyns [17] replaced the lesioning procedures, with their irreversible effects, by stimulation at high frequencies, which has a blocking effect that is fully reversible. Bilateral DBS of the anterior limb of the internal capsule by Nuttin and Cosyns yielded significant improvement of OCD symptoms in four patients, but unusually high stimulation amplitudes had to be used, which caused high energy consumption of the impulse generators and consequently their frequent servicing.

The fact that the distal lead of the electrodes used (Medtronic, Minneapolis, USA) is placed into the ventral edge of the internal capsule, where it abuts the nucleus accumbens, and the high stimulation amplitudes make functional blocking of accumbens activity highly probable.

The extensive experience of the Karolinska group [15], as well as our own observations with anterior capsulotomy for treating OCD and anxiety disorders, show that lesioning of the ventrocaudal part of the internal capsule is decisive for successful treatment.

Similar conclusions have been made by Rasmussen and Greenberg (2002, personal communication) for gamma-capsulotomy. Thermocoagulation and radiation necrosis in the ventral edge of the internal capsule are likely to affect the nucleus accumbens as well, including its shell region.

Target Selection and Clinical Findings

Considering the central position of the nucleus accumbens between the amygdaloid complex, basal ganglia, mediodorsal thalamic nucleus, and prefrontal cortex, all of which are involved in the pathophysiology of anxiety disorders [22] and OCD [20], the beneficial clinical effects of anterior capsulotomy might well be caused by blocking of the amygdaloid-basal ganglia-prefrontal circuitry at the level of the shell region of the nucleus accumbens, rather than by blocking of the fiber tracts in the internal capsule.

These considerations prompted us to modify the electrode track for DBS. Instead of targeting the anterior limb of the internal capsule alone, we implanted the electrode in a way that permitted stimulation of the ventral part of the anterior limb of the internal capsule as well as of the shell region of the nucleus accumbens with the same electrode and selective stimulation of these structures.

For the first patient treated in the pilot series we implanted the DBS electrodes bilaterally. Alternating activation of various contact combinations of the right, left, or both electrodes was performed during a testing period of several weeks following electrode implantation. We observed that bipolar stimulation over the two distal leads of the right electrode (0 negative, 1 positive), which were placed within the right nucleus accumbens, yielded significant reduction in symptoms. Bilateral stimulation did not improve the effects. Activation of leads placed in the internal capsule had not been effective. Consequently, in the following three patients we implanted the electrode only unilaterally into the right nucleus accumbens (◘ Fig. 16.1).

The DBS electrode implantation is performed stereotactically. The positioning of the electrode is

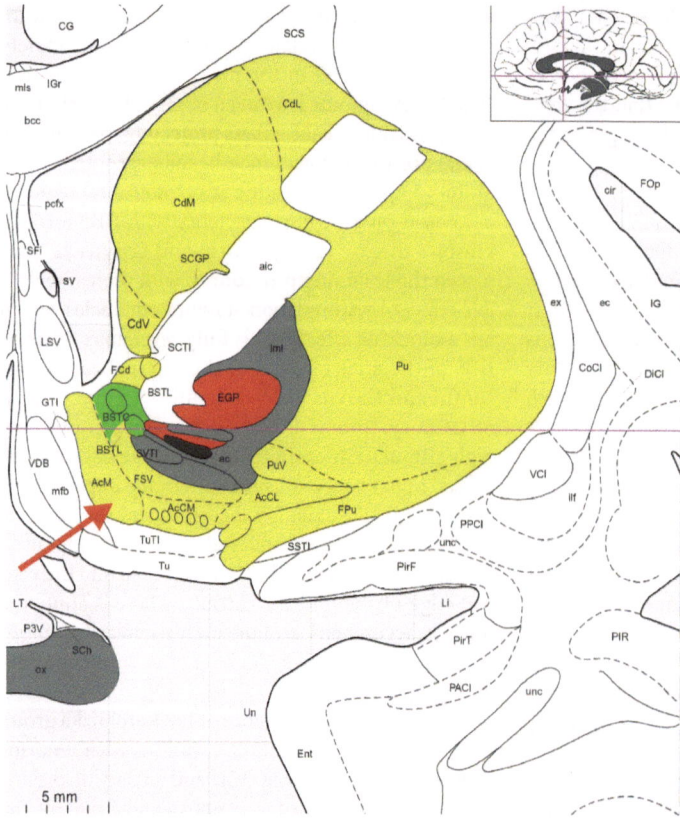

Fig. 16.1. Frontal section of the human brain at the level of the target point in the ventro-medio-caudal part of the nucleus accumbens. Arrow indicates target point. Coordinates: 2.5 mm rostral anterior border of AC, 6.5 mm lateral of midline, 4.5 mm ventral AC. (Mai et al. 1997)

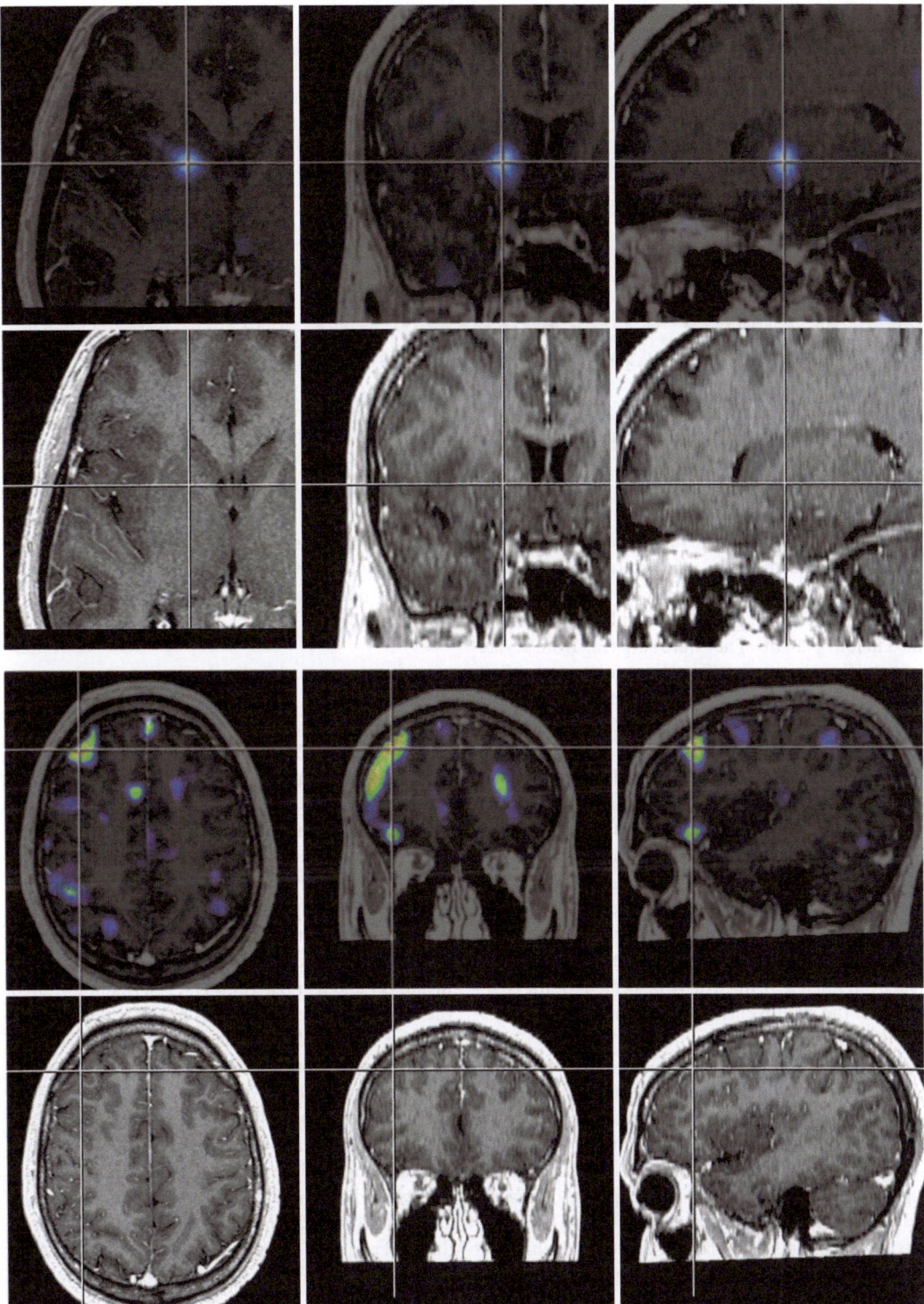

◼ Fig. 16.2. Upper image: region of neuronal inhibition as measured by 15O-H2O-PET superimposed onto the preoperative MR image. Lower image: frontal cortex shows neuronal activation as a result of DBS in the ipsilateral (right) shell of the nucleus accumbens

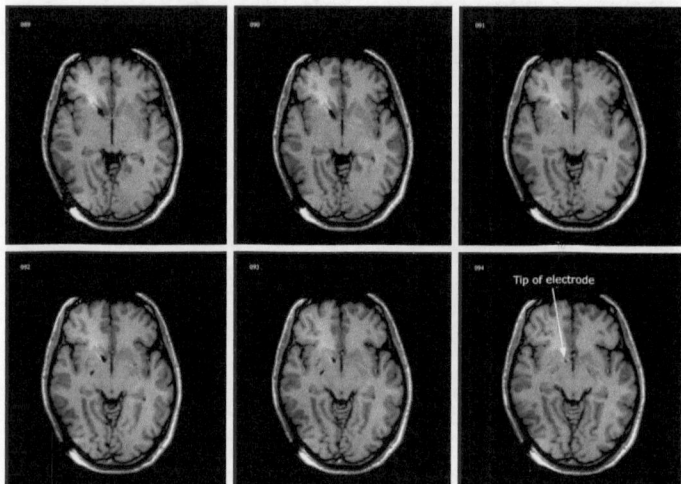

Fig. 16.3. Postoperative T1-weighted MRI depicts exact electrode placement in the desired target area. Note the dark artifact surrounding the electrode, which is due to disturbance of the local magnetic field by the electrode and not caused by tissue damage

verified by intraoperative X-ray. Target point and trajectory have been assessed using stereotactic MR and CT imaging (■ Figs. 16.2 and 16.3).

Unilateral DBS of the right accumbens was performed in four patients. They were suffering from severe anxiety disorders and OCD and no longer responding to medical treatment and psychotherapy. DBS treatment, applied with permanent pulse-train stimulation (square-wave impulses of 90 μs duration, 130 Hz, and amplitudes between 2 and 6.5 V) yielded nearly total recovery from both anxiety and OCD symptoms without any side effects in three of four patients with follow-up periods of 24–30 months. Clinical improvement occurred a few days to several weeks after the beginning of DBS. In the fourth patient no beneficial effect was achieved. A recently performed MRI investigation in that patient revealed that the target area had been missed owing to a displacement of the electrode in the caudoventral direction, which explains the therapeutic failure.

In one patient a 15O-H2O-PET study in conditions "stimulation on vs. stimulation off" was performed. High-frequency stimulation of the shell of the nucleus accumbens inhibited the activity of the ipsilateral dorsolateral rostral putamen but activated the right dorsolateral prefrontal and cingulate cortex.

One patient had a severe relapse of OCD and anxiety symptoms 30 months after electrode implantation and permanent DBS. Following exchange of the pacemaker, the patient recovered very well and was able to return to "normal" life 3 days later.

This article focuses on a possible rationale for DBS in the nucleus accumbens in the treatment of anxiety and obsessive-compulsive disorder. A detailed presentation of the clinical and PET data (not yet published) would exceed the scope of the present communication.

Discussion

The significant improvement of symptoms due to severe anxiety and OC disorders obtained with unilateral high-frequency stimulation of the shell of the right n. accumbens indicates a major role for this nucleus as a central relay between the amygdaloid complex, the basal ganglia, the mediodorsal thalamus, and the prefrontal cortex. The amygdaloid complex, especially the lateral nucleus, is well-known to be involved in anxiety and fear reactions [11, 22]. Disinhibition of the lateral amygdaloid nucleus has been shown to be decisive for the development of Pavlovian learned fear, depending on a deficiency of gastrin-releasing peptide receptors on GABA-ergic interneurons, which in turn causes disinhibition of principal neurons in the lateral-amygdaloid nucleus [22].

Pathological information flow from the lateral amygdaloid nucleus can be propagated to basolateral and central amygdaloid nuclei, to finally converge in the shell region of the accumbens via both the ventral amygdalofugal pathway and the extended amygdala. The shell region could thus represent a "bottleneck" for impulse propagation from the amygdaloid complex to the basal ganglia, mediodorsal thalamus, and prefrontal cortex, areas involved in the pathophysiology of OCD, as shown with functional imaging [22]. This was supported by our PET data, briefly described above.

The PET findings seemed to correlate well with the symptoms of the patient, which were aggravated in the stimulation-off period and relieved in the stimulation-on period. A placebo effect cannot be excluded because the investigated patient had not been blinded.

The good results in our anxiety and OCD patients might be explained by blocking of this hypothetical pathological impulse flow through chronic high-frequency stimulation of the shell of the N. accumbens. It is noteworthy that unilateral stimulation of the right n. accumbens was sufficient. This finding is in line with the results of Lippitz et al. [12], who found that capsulotomies in the right hemisphere were decisive for a favorable therapeutic outcome.

Inputs from the amygdaloid complex to the nucleus accumbens "gate" both fronto-striatal and hippocampo-striatal circuitry [5, 16, 19]. It might thus be speculated that a dysfunction of the nucleus accumbens, resulting in an impairment to adequately modulate amygdalo-basal ganglia-prefrontal circuitry, might be at the origin of anxiety disorders and OCD.

References

1. Alheid GF, Beltramino CA, de Olmos JS, Forbes MS, Swanson DJ, Heimer L (1998) The neuronal organization of the supracapsular part of the stria terminalis in the rat: the dorsal component of the extended amygdala. Neuroscience 84: 967–996
2. Benabid AL, Pollak P, Gervason C, Hoffmann D, Gao DM, Hommel M, Perret JE, De Rougemont J (1991) Long-term suppression of tremor by chronic stimulation of ventral intermediate thalamic nucleus. Lancet 337: 402–406
3. Blond S, Caparros-Lefebvre D, Parker F, Assaker R, Petit H, Guieu JD, Christiaens JL (1992) Control of tremor and involuntary movement disorders by chronic stereotactic stimulation of the ventral intermediate thalamic nucleus. J Neurosurg 77: 62–68
4. de Olmos J, Heimer L (1999) The concepts of the ventral striatopallidal system and extended amygdala. Ann NY Acad Sci 877: 1–31
5. Grace AA (1993) Cortical regulation of subcortical dopamine systems and its possible relevance to schizophrenia. J Neural Transm 9: 111–134
6. Haber SN (2003) The basal ganglia: parallel and integrative networks. J Chem Neuroanat this volume
7. Haber SN, Fudge JL, McFarland NR (2000) Striatonigrostriatal pathways in primates form an ascending spiral from the shell to the dorsolateral striatum. J Neurosci 20: 2369–2382
8. Heimer L (2000) Basal forebrain in the context of schizophrenia. Brain Res Rev 31: 205–235
9. Heimer L, Harlan RE, Alheid GF, Garcia M, de Olmos J (1997) Substantia innominata: a notion which impedes clinical-anatomical correlations in neuropsychiatric disorders. Neuroscience 76: 957–1006
10. Jenike MA (1998) Neurosurgical treatment of obsessive-compulsive disorder. Br J. Psychiatry [Suppl] 35: 79–90
11. Le Doux JE (2000) Emotion circuits in the brain. Annu Rev Neurosci 23: 155–184
12. Lippitz BE, Mindus P, Meyerson BA, Kihlstrom L, Lindqvist C (1999) Lesion topography and outcome after thermo-capsulotomy or gamma knife capsulotomy for obsessive-compulsive disorder: relevance of the right hemisphere. Neurosurgery 44: 452–458
13. Mai JK, Assheuer J, Paxinos G (1997) Atlas of the human brain. Academic Press, San Diego
14. Matthysse S (1973) Antipsychotic drug actions: a clue to the neuropathology of schizophrenia? Fed Proc 32: 200–205
15. Meyerson BA (1998) Neurosurgical treatment of mental disorders: introduction and indications. In: Gildenberg PL, Tasker RR (eds) Textbook of stereotactic and functional neurosurgery. McGraw-Hill, New York, pp 1953–1963
16. Mulder AB, Hodenpijl MG, Lopes da Silva FH (1998) Electrophysiology of the hippocampal and amygdaloid projections to the nucleus accumbens of the rat: convergence, segregation, and interaction of inputs. J Neurosci 18: 5059–5102
17. Nuttin B, Cosyns P (1999) Electrical stimulation in anterior limbs of internal capsules in patients with obsessive-compulsive disorder. Lancet 354: 1526
18. Rauch SL, Dougherty DD, Cosgrove JR, Cassem EH, Alpert NM, Price BH, Nierenberg AA, Mayberg HS, Baer L, Jenike MA, Fischman AJ (2001) Cerebral metabolic correlates as potential predictors of response to anterior cingulotomy for obsessive compulsive disorder. Biol Psychiatry 50: 659–667
19. Rosenkranz JA, Grace AA (1999) Modulation of basolateral amygdala neuronal firing and afferent drive by dopamine receptor activation in vivo. J Neurosci 15: 11027–11039
20. Saxena S, Rauch SL (2000) Functional neuroimaging and the neuroanatomy of obsessive-compulsive disorder. Psychiatr Clin North Am 23: 563–586
21. Schuurman PR, Bosch DA, Bossuyt PM, Bonsel GJ, van Someren EJ, de Bie RM, et al (2000) A comparison of continuous thalamic stimulation and thalamotomy for supression of severe tremor. N Engl J Med 342: 461–468
22. Shumyatsky GP, Tsvetkov E, Malleret G, Vronskaya S, Hatton M, Hampton L, Battey JF, Dulac C, Kandel ER, Bolshakov VG (2002) Identification of a signaling network in lateral nucleus of amygdala important for inhibiting memory specifically related to learned fear. Cell 111: 905–918
23. Stevens JR (1973) An anatomy of schizophrenia? Arch Gen Psychiatry 29: 177–189
24. Volkmann J, Sturm V, Weiss P, Kappler J, Voges J, Koulousakis A, Lehrke R, Hefter H, Freund HJ (1998) Bilateral high-frequency stimulation of the internal globus pallidus in advanced Parkinson's disease. Ann Neurol 44: 953–961

Stimulation of the Posterior Hypothalamus for Medically Intractable Impulsive and Violent Behavior

A. Franzini, C. M. P. Ferroli, O. Bugiani, G. Broggi

Introduction

In 1970, Sano [1] reported that the lesion of the posterior hypothalamus by means of stereotactic radiofrequency was effective in treating disruptive and aggressive behavior. More recently, positron emission tomography (PET) has shown activation of ipsilateral posterior-inferior hypothalamic gray matter during attacks of chronic cluster headache (CH) [2, 3]. The high frequency electrical chronic stimulation (HFS) of the hypothalamic area has been found to be effective in CH [4–6], a condition in which violent behavior and psychomotor agitation may develop during pain attacks [7, 8].

High-dose neuroleptics are employed to control aggressive and acting out patterns but in some patients this pharmacological treatment is ineffective and associated with severe extrapyramidal side effects [9, 10].

Two male patients came to our observation, aged 36 and 37 years, respectively, suffering from mental retardation with aggressive and disruptive behavior, resistant to pharmacological treatment. The first patient also complained of grand mal seizures and the EEG showed the presence of many multifocal spikes. He was unable to speak and could only modulate inarticulate sounds. He had been under treatment with neuroleptics and anti-epileptics and lab examinations showed signs of liver function failure, probably re-

lated to the prolonged administration of elevated doses of the medicines. This made it compulsory to resort to an alternative treatment.

The second patient showed labio-palatoschisis, chorioretinitis and moderate oligophrenia probably attributable to congenital toxoplasmosis. At the age of 17 he had been admitted to a psychiatric institution. Many attempts to reintroduce him to a domestic environment had failed. Treatments with neuroleptics, anti-epileptics and benzodiazepines were unsuccessful in controlling aggressiveness.

Methods

After the informed consent was obtained from the parents, under general anesthesia both patients underwent stereotactic bilateral electrode implant in the medial portion of the posterior hypothalamus. Two 4-contact electrodes (Quad 3387, Medtronic Inc.) were inserted through a 3 mm, coronal paramedian twist-drill hole at the appropriate coordinates. Postoperative stereotactic computerized tomography (CT) was merged with pre-operative magnetic resonance imaging (MRI) to confirm the correct placement of electrodes. Two pulse generators (Soletra, Medtronic) were placed in the subclavicular region and connected to the brain electrodes. The day after surgery, bilateral continuous monopolar 185 Hz, 1 volt, 60 μs electrical stimulation was started. No side effects occurred.

Results

Neuroleptic administration was interrupted in the first patient 2 weeks after the start of stimulation. The patient appeared calmer and more cooperative and a few weeks later he was able to stand and walk. One year later these effects were still present: the patient had regained a normal circadian rhythm, was able to take care of himself and to undergo rehabilitation. His family relationships and social interests markedly improved. Epileptic seizures decreased from 7–10 to 4–7 per day.

In the second patient, aggressive behavior, including acting-out, completely disappeared and dosage of neuroleptics could be reduced. Three months later his psychiatric condition was stable, and he was transferred to a specialized center for occupational therapy.

Conclusions and Implications for Clinical Practice Today

Deep Brain Stimulation (DBS) is thought to act through the functional inhibition of targeted areas produced by HFS, an effect similar to the one reported by Sano with radiofrequency lesions [1]. The result is an attenuation of behavioral abnormalities of patients with mental retardation secondary to brain damage. Hypothalamus is a core structure of the limbic system which connects hippocampus, involved in learning and memory, and amygdala, associated with emotions, affiliative behavior and with autonomic and endocrine functions. Hypothalamus is also connected to the orbito-frontal cortex via the amygdala and the limbic thalamus. Stimulation of the posterior hypothalamus has been shown to be effective in patients with CH without producing behavioral effects [5], while in the 2 cases reported in this paper, it caused disruptive behavior to disappear, at the same time

markedly improving social relationships and quality of life of the subjects. This seems to suggest that the neurostimulation of the same brain target may induce different effects according to the different existing clinical conditions.

Future Directions

HFS of hypothalamus appears to be a clinically and ethically acceptable technique in patients with aggressive behavior when conservative treatments are not applicable and pharmacological treatment is ineffective or causes important, unacceptable side effects.

References

1. Sano K, Mayanagi Y, Sekino H, Ogashiwa M, Ishijima B. Results of stimulation and destruction of the posterior hypothalamus in man. J Neurosurg 1970; 33: 689–707
2. Goadsby PJ. Neuroimaging in headache. Appl Neurophysiol 1982; 45: 136–142
3. May A, Bahra A, Buchel C, Frackowiak RS, Goadsby PJ. PET and MRA findings in cluster headache and MRA in experimental pain. Neurology 2000; 55: 1328–1335
4. Leone M, Franzini A, Bussone G. Stereotactic stimulation of posterior hypothalamic gray matter in a patient with intractable cluster headache. N Engl J Med 2001; 345: 1428–1429
5. Franzini A, Ferroli P, Leone M, Broggi G. Stimulation of the posterior hypothalamus for treatment of chronic intractable cluster headaches: First reported series. Neurosurgery 2003; 52: 1095–1099
6. Leone M, Franzini A, Broggi G, May A, Bussone G. Long-term follow-up of bilateral hypothalamic stimulation for intractable cluster headache. Brain 2004; 127: 2259–2264
7. Torelli P, Manzoni GC. Pain and behaviour in cluster headache. A prospective study and review of the literature. Funct Neurol 2003; 18: 205–210
8. Manzoni GC, Terzano MG, Bono G, Micieli G, Martucci N, Nappi G. Cluster headache: Clinical findings in 180 patients. Cephalalgia 1983; 3: 21–30
9. Bauermeister JJ, Canino G, Bird H. Epidemiology of disruptive behavior disorders. Child Adolesc Psychiatry Clin North Am 1994; 3: 177–194
10. Dosen A. Diagnosis and treatment of psychiatric and behavioral disorders in mentally retarded individuals: The state of the art. J Intellect Disabil Res 1993; 37: 1–7

Gilles de la Tourette's Syndrome: A Movement Disorder

C. van der Linden

Gilles de la Tourette's syndrome (TS) is a genetic disorder with onset in childhood and characterized by phonic and motor tics. Typically, the disorder starts around the age of seven with simple motor tics, such as forceful eye blinking, followed by simple vocal tics such as grunting noices or throat clearing. Subsequently, the tics become more complex, sometimes resembling purposeful movements. Vocal tics may become more complex, sometimes leading to coprolalia, i.e. the use of obscenic words. Boys are four times more effected than girls. The mode of inheritance is unknown, but is it believed to be autosomal dominant with variable penetrance. Behavioral disturbances are associated with TS. Obsessive-compulsive behavior (OCB), attention-deficit/hyperactivity disorder (ADHD) and loss of impulse control are co-morbid psychiatric disorders. The etiology and pathogenesis are unknown. Several lines of research are indicative of involvement of the basal ganglia in the generation of tics. The basal ganglia are probably the key structure in the pathophysiology of tics in TS. Various circuits have been described in which activity originating from the frontal cortex leads back to the frontal cortex via the basal ganglia and thalamus, the so-called cortico-striato-thalamo-cortical loops [17, 18]. The various loops run parallel to each other and each have their own function, varying from a sensorimotor integrative to a more complicated cognitive and behavioral function. These cognitive and behavioral loops probably play an important role in the pathogenesis of tics in TS. All these circuits run through the internal pallidum (GPi), which serves as the major output structure of the basal ganglia. Via various thalamic nuclei, including the ventrolateral nucleus and the more median located nuclei such as the centromedian and parafascicular nuclei, the loops project back to the frontal cortex. Within the basal ganglia, two major pathways have been identified, the direct and the indirect, which connects the input and output of the basal ganglia (◨ Fig. 18.1). Using this simplified basal ganglia model one can hypothesize the pathogenesis of the various hypokinetic and hyperkinetic movement disorders. In TS, a typical hyperkinetic disorder, an altered modulation of the striatum giving rise to an increased inhibition of the GPi and disinhibition of the thalamo-cortical projection may be involved in the pathogenesis. This altered inhibition of the GPi may be induced by abnormal activity originating from the pre-frontal cortex [19]. Moreover, animal experiments show that stereotypic behavior could be induced by abnormal activity from the striatum [20, 21]. Using PET en SPECT technology, in-vivo studies reveal a disturbance of both the presynaptic and postsynaptic striatal dopamine receptors in patients with TS [22, 23]. Taking all those observations into account, altered modulation of the dopamine input seems important in the genesis of tics. This had been hypothesized for decades by the notion that dopamine antagonists have a favorable effect in controlling tics [24]. In addition, lesions in the mesencephalon, in which there is a large concentration of dopamine-containing neurons, have been described to cause tics (25).

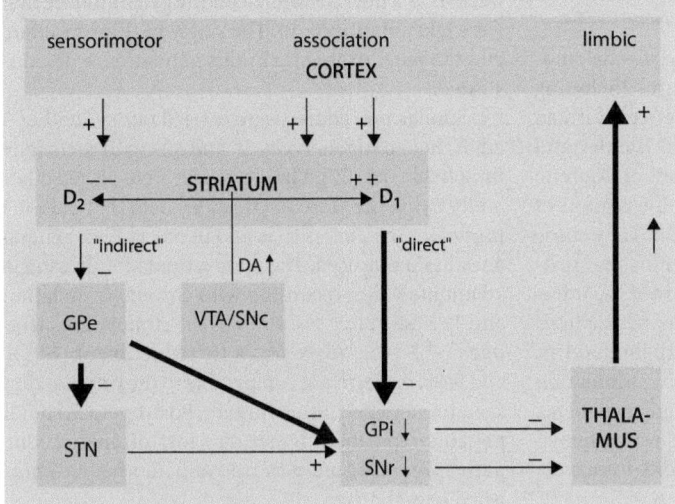

□ **Fig. 18.1.** Schematic representation of the direct and indirect pathways. In Gilles de la Tourette syndrome a hyperdopminergic state leads to inhibition of the indirect pathways and stimulation of the direct pathways resulting in inhibition of the SNr en GPi. Inhibition of the SNr/GPi complex facilitates the thalamocortical projections (Bruggeman 2001, with permission)

Treatment

The treatment of Gilles de la Tourette syndrome (TS) needs different approaches. In the first place, it is important to give detailed information to the patient, parents, school or workplace. Generally, this information is sufficient for TS patients to cope with motor and vocal tics. Secondly, if tics interfere with social and professional activities pharmacological treatment may be necessary. Several classes of anti-tic medication are available, in particular the alpha-adrenergic drug clonidine and neuroleptics such as haloperidol and pimozide and more recently atypical neuroleptics such as risperidone [1] Thirdly, recent studies have shown that behavioral therapy may, at least in part, control the tics [2]. Most patients with TS will have a significant reduction of tics by the time they reach adulthood (3). However, a small portion of TS patients continue to have bothersome tics with interference of both social and professional life despite adequate pharmacological treatment. In those patients brain surgery has been employed since the early 60s. Detailed data on the short and long term results are lacking and serious side effects have limited their general use.

Initially, neurosurgical procedures consisted of the destruction of various brain tissues on the basis of empirical data. Most of these reported patients were operated, because of associated psychiatric disturbances, in particular obsessive compulsive disorder (OCD) [4]. Frequently, the tics were not responsive to the surgical procedure. If tics were reduced

by the surgery, it was unclear which target was responsible for the reduction of tics, because of the lack of selective lesions [4]. In addition, the lesions were very large. Regions in the vicinity of the presumed target could have contributed to the reduction of the tics. □ Table 18.1 shows an overview of the various presumed targets of recent publications. All authors report serious adverse events in several of the operated patients [5–13].

Because of the serious morbidity, neurosurgery as a treatment option in TS was generally abandoned by most specialized TS centers. Due to the refinement of the stereotactic technique and the safe procedure of deep brain stimulation (DBS) in other movement disorders, such as Parkinson's disease, tremors and dystonia [14–16], the neurosurgical treatment has gained new attention.

□ **Table 18.1.** Brain targets, which have been used for lesioning for the treatment of tics in TS. References in parentheses

Frontal cortex [5]
Gyrus cinguli anterior [6]
Limbic area [7, 8]
Thalamus [9, 10]
Infrathalamic area [11, 12]
Zona incerta [12]
Nucleus dentatus [13]

Neurosurgery

Before the description of the cortico-striato-thalomo-cortical loops, neurosurgical procedures including thalamic lesioning and leucotomies were performed on an empirical basis. It was not until Hassler and Dieckman described their surgical cases of Tourette patients with intractable tics that specific regions of the brain were targeted for lesioning [10]. They chose the medial thalamus as the main target for the treatment of tics, whereas Babel et al. [12] added the infra-thalamic region as a target on the basis of neurophysiological studies in TS. Side effects have limited this procedure. In a recent study, deep brain stimulation (DBS) was shown to be safer than lesioning in patients treated for tremor [15]. Therefore, DBS was assumed to be the neurosurgical treatment of choice for intractable tics.

The principal of electrical stimulation of a target is believed to be similar to lesioning, since both methods inhibit the activity of the target. Vandewalle et al. reported on bilateral medial thalamic stimulation in a 37 year old male TS patient resistant to conventional tic therapy [26]. The target was chosen on the basis of the reported lesioning by Hassler and Dieckman in the medial thalamus [10]. Because of the multitude of lesions in Hassler's cases, the quadripolar electrode was placed in such a direction that many of the reported lesions by Hassler and Dieckman could theoretically be involved in the stimulated area. The aforementioned cortico-striato-thalamo-cortical loops, including the medial thalamus with the centromedian

nucleus as a possible source for the generation of tics, were taken into account. The safety of this procedure and the relief of the tics in this patient were demonstrated.

Similar procedures were carried out in another 2 adult male patients (ages 28 and 42) with medically intractable tics [27]. The electrodes were placed bilaterally using the stereotactic approach. In 2 patients propofol tuned anesthesia and in one patient general anesthesia was used. Tics were scored blindly using a 20 minute video-recording with chronic stimulation and 12 hours after cessation of the stimulation. After one and 5 years there was a tic reduction of 72,3% and 90%, respectively, comparable to the tic reduction seen immediately after surgery. Post-operatively, all patients reported a temporary loss of energy. One patient reported an increased and another a diminished sexual drive.

In a 27 year old male TS patient, medial thalamic stimulation was compared to bilateral internal pallidal stimulation, by placing one quadripolar electrode in each target bilaterally, thereby implanting a total of four intracerebral electrodes [28]. The stimulation of the internal pallidum appeared to be more effective than the medial thalamus in reducing the tics. The ventrolateral part of the internal pallidum, i.e. the target used for the treatment of dystonia and for some patients with Parkinson's disease was selected (◘ Fig. 18.2).

Based on these results, the target to be used for chronic stimulation in the control of intractable tics in TS remains to be determined. All selected patients

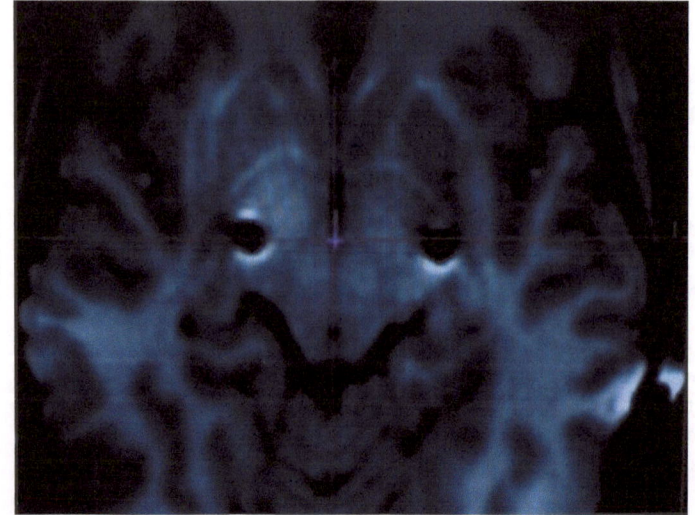

◘ **Fig. 18.2.** Post-operative MR T1 axial image through the basal ganglia in the patient with bilateral pallidal stimulation. Note the artefact from the electrodes located in the ventrolateral part of the internal pallidum

had low levels of comorbidity, because the effect of chronic stimulation on associated symptoms such as obsessive-compulsive behavior (OCB), attention deficit-hyperactivity disorder (ADHD) or loss of impulse control is unclear.

Conclusion

TS is a neurological disorder in childhood, frequently "self-limiting", not requiring any treatment into adulthood. Counseling and behavioral therapy may suffice for the treatment of tics and comorbidity. Pharmacological therapy is in some cases necessary and in most cases effective. In rare instances in patients with TS in adulthood, who do not respond to various forms of non-invasive treatment, neurosurgical therapy may be effective in suppressing tics. The lesioning technique of brain targets is obsolete due to the serious adverse events reported in most of the cases. The deep brain stimulation technique appears promising due to its efficacy and safety in the four described cases (◘ Fig. 18.3).

However, it remains to be determined which is the preferred target for optimal tic control. Based on the current understanding of the pathophysiology of the Gilles de la Tourette syndrome, both targets, i.e. medial thalamus and GPi, can theoretically be considered. Prospective studies are underway to evaluate the effect of the different targets on the tics and its efficacy and safety.

References

1. Bruggeman R, Van der Linden C, Op den Velde W et al. A comparative double-blind parallel group study of risperidone versus pimozide in Gilles de la Tourette's syndrome. J Clin Psychiatry 2001; 62: 50–56
2. Hoogduin CAL, de Haan E, Cath DC, van de Wetering BJM. Gedragstherapie. In: Buitelaar JK, van de Wetering BJM (eds) Syndroom van Gilles de la Tourette: Een leidraad voor diagnostiek en behandeling. Assen, Van Gorkum 1994, pp 61–67
3. Leckman JF, Zhang H, Vitale A et al. Course of tic severity in Tourette's syndrome: the first two decades. Pediatrics 1998; 102: 14–19
4. Rauch SL, Bear L, Cosgrove and Jenike M. Neurosurgical treatment of Tourette's syndrome: a critical review. Comp Psychiatry 1995; 36: 141–156
5. Baker EFW. Gilles de la Tourette syndrome treated by medial frontal leucotomy. Can Med Assoc J 1962; 86: 746–747
6. Kurlan R, Kersun J, Ballantine T, Caine ED. Neurosurgical treatment of severe obsessive-compulsive disorder associated with Tourette's syndrome. Mov Disord 1990; 5: 152–155
7. Sawle GV, Lees AJ, Hymas NF, Brooks DJ, Frackowiak RSJ. The metabolic effects of limbic leucotomy in Gilles de la Tourette syndrome. J Neurol Neurosurg Psychiatry 1993; 56: 1016–1019
8. Robertson M, Doran M, Trimble M, Lees AJ. The treatment of Gilles de la Tourette syndrome by limbic leucotomy. J Neurol Neurosurg Psychiatry 1990; 53: 691–694
9. Diviitis E, D'Errico, Cerillo A. Stereotactic surgery in Gilles de la Tourette syndrome. Acta Neurochir (suppl)1977; 24: 73
10. Hassler R, Dieckmann G. Traitement stéréotaxique des tics et cris inarticulés ou coprolalique considérés comme phénoméne d'obsession motrice au cours de la maladie de Gilles de la Tourette. Rev Neurol (Paris) 1970; 123: 89–100

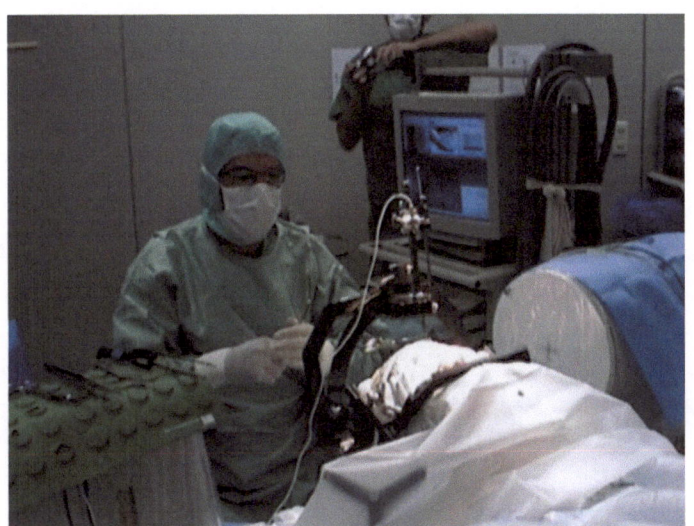

Fig. 18.3. Operating theatre with CRW frame fixed to the patient's skull and test electrode in place

11. Leckman JF, de Lotbinière AJ, Marek K et al. Severe disturbances in speech, swallowing, and gait following stereotactic infrathalamic lesions in Gilles de la Tourette's syndrome. Neurol 1993; 43: 890–894

12. Babel TB, Warnke PC, Ostertag CB. Immediate and long term
outcome after infrathalamic and thalamic lesioning for
intractable Tourette's syndrome. J Neurol Neurosurg Psychiatry 2001; 70: 666–671

13. Nadvornik P, Sramka M, Lisy L, Svicka J. Experiences with
dentatomy. Confin Neurol 1972; 34: 320–324

14. Limousin P, Krack P, Pollack P, et al. Electrical stimulation of
the subthalamic nucleus in advanced Parkinson's disease.
New Engl J Med 1998; 330:1105–1111

15 Schuurman PR, Bosch BA, Bossuyt PMN et al. A comparison
of continuous thalamic stimulation and thalamotomy for
suppression of severe tremor. New Eng J Med 2000; 342:
461–468

16 Coubes P, Roubertie A, Vayssiere N, Hemm S, Echenne B.
Treatment of DYT-1 generalized dystonia by stimulation of
the internal globus pallidus. Lancet 2000; 355: 2220–2221

17. Alexander GE, Crutcher MD. Functional architecture of basal ganglia circuits: neural substrates of parallel processing.
Trends Neurosci 1990; 13: 266–271

18. Groenewegen HJ. Bewegen: de rol van de basale ganglia.
In Wolters EC, van Laar T (eds) Bewegingsstoornissen.
Amsterdam, VU uitgeverij 2002: 3–34

19. Leckman JF. Tourette's syndrome. Lancet 2002; 360:
1577–1586

20. Canales JJ, Graybriel AM. A measure of striatal functions
predicts motor stereotypy. Nat Neurosci 2000; 3: 377–383

21. Graybiel AM, Canalis JJ. The neurobiology of repetitive behaviors: clues to the neurobiology of Tourette syndrome.
Adv Neurol 2001; 85: 123–131

22. Ernst M, Zametkin AJ, Jons PH et al. High presynaptic
dopaminergic activity in children with Tourette's disorder.
J Am Acad Child Adolesc Psychiatry 1999; 38: 86–93

23. Wolf SS, Jones DW, Knable MB et al. Tourette syndrome:
prediction of phenotypic variation in monozygotic twin by
caudate nucleus D2 receptor binding. Science 1996; 273:
1225–1227

24. Challas G and Brauer W. Tourette's disease: relief of symptoms with R 1625. Am J Psych 1963; 120: 283–284

25. Van der Linden C, Bruggeman R, Pengel J. Gilles de la
Tourette syndrome in a patient with a mesencephalic lesion. Mov Disorders 1994; 9: 252

26. Vandewalle V, van der Linden C, Caemaert J, Groenewegen
HJ: Stereotactic treatment of Gilles de la Tourette syndrome
by high frequency stimulation of thalamus. Lancet 1999;
353: 724

27. van der Linden C, Colle H, Vandewalle V, Alessi G, Rijckaert
D, De Waele L. Chronic bilateral medial thalamic (MT) stimulation for the treatment of tics in 3 adult patients with
Gilles de la Tourette syndrome. Movement Disorders 2002;
188 (Suppl)

28. van der Linden C, Colle H, Vandewalle V, Alessi G, Rijckaert
D, De Waele L Successful treatment of tics with chronic bilateral internal pallidum (GPi) stimulation in a 27 year old
male patient with Gilles de la Tourette's syndrome. Movement Disorders 2002; 341 (Suppl)

18

Advances in Gilles de la Tourette Syndrome: Preliminary Results in a Cohort of 10 Patients Treated with DBS

M. Porta, D. Servello, M. Sassi, A. Brambilla

Introduction

Tourette syndrome (TS) is a disorder characterized according to DSM IV [1] by:
- presence, simultaneously or not, of motor and vocal tics of a duration longer than 1 year, in a continuous or intermittent manner;
- variability during time in rate, intensity and patterns of tics;
- onset of disturbances before the age of 18;
- exclusion of secondary tics, attributable to psychotropic drugs, infectious diseases, intoxications, altered CNS development, CNS trauma, stroke, use of neurotoxic substances such as cocaine, etc.; need of a diagnosis supported by an adequate video/cinematographic documentation evaluated by an expert;
- psychosocial impact.

Jankovic [2] has distinguished the following types of tics:
- simple motor (clonic, dystonic, tonic)
- complex motor (apparently purposeless, inadequate, finalistic)
- simple vocal (cough, grunting)
- complex vocal (echolalia, palilalia, often coprolalia).

Jankovic [3] has also described physiology of movements as follows:

- voluntary,
- unvoluntary,
- involuntary,
- automatic.

Besides voluntary movements, during all the other 3 types, tics can be produced.

Tics are purposeless movements which can be voluntarily and temporarily suppressed but which suddenly reappear in bursts, with a variable time course and affecting several muscle groups such as the respiratory, laryngeal, pharyngeal, oral and nasal muscles. Tics tend to disappear when the subject is performing mental or physical activities, including sexual intercourse. Paradoxically, the clinical picture deteriorates when the subject is relaxed or resting. During sleep, pathological movements may disappear but more frequently they tend to persist.

For the diagnosis, for a better assessment of symptoms and progress of the disease, a number of evaluation tools have been proposed such as the Yale Global Tic Severity Scale (YGTSS), the Tourette Syndrome Questionnaire (TSQ), the Tourette Syndrome List (TSL), the Unified Tourette Syndrome Rating Scale and a rating score for the video assessment of the disease which takes into account the anatomic location of tics, types of vocal tics, the severity and the frequency of tics. It has been observed that patients with TS often complain also of Obsessive Compulsive Behavior (OCB) and/or of Attention Deficit Hyperactive Disorder (ADHD). Anxiety is also present in these patients as Self Injury behavior (SIB), originally described also by Gilles de la Tourette. We used the Robertson-Cohen Classification of Tourette Syndrome, differentiating patients with simple, full-blown, or plus form of the disease. Patients refractory to standard and innovative drug- and psycho-, conservative therapies were considered in a distinct subgroup labeled "resistant". Patients belonging to this group were considered eligible for surgery.

TS is recognized to be more common than was previously reported with prevalence figures from recent studies [4] of between 0,4% and 1,76% of youngsters between the ages of 5 and 18 years.

Physiopathology and Pathogenesis

It has not yet been possible to identify histopathological lesions responsible for the onset of TS: It has been postulated that there is the involvement of the caudate nucleus and of the lower area of pre-frontal cortex, similar to what has been supposed for OCD and ADHD [6, 7]. Volumetric MRI [7] has shown asymmetry in basal ganglia while functional MRI [8] has shown that during suppression of tics, subcortical extrapyramidal areas are hypoactive while the activity of inhibitory cortical areas is increased.

By means of PET with 18-fluoro-2-deoxy-D-glucose [9] two metabolic patterns have been identified: the first characterized by increased activity in lateral pre-motor and supplementary motor areas, the second characterized by decreased activity of the caudate nucleus and of thalamic areas belonging to the projection circuit limbic system-basal ganglia-thalamus-cortex.

Biochemical data point to a reduction of cortical cyclic AMP [6, 10] and to alterations in dopamine uptake [11] but also changes in neurotransmission of GABA, neuropeptides, acetylcholine and N-methyl-D-aspartate appear to be present. On the basis of behavioral patterns, Pinelli et al., in collaboration with Porta [12], have clinically differentiated 2 types of patients with TS, one with predominant disorder of the serotoninergic system and the other with predominant disorder of the dopaminergic system.

Multiple factors have been considered to play a role in the onset of TS, such as heredity, age, gender and environment. Also infections have been considered as possible causative factors, particularly infections from beta-hemolytic Streptococcus, but also Borrelia and Herpes simplex have been considered

possible agents triggering the onset of the syndrome. Alsobrook and Pauls in 2002 demonstrated heritability in 3 out of 4 factors, among which coprolalia, aggressivity, simple motor and phonic tics. More recently, Abelson et al. (2005) reported the association of TS with the gene SLITRK1 on chromosome 13q31.1 in a small number of individuals with TS, which is the first indication of an actual gene being involved in some cases of TS.

Management

The management of patients with TS presents several difficulties. The first problem is to decide which type of patient should be treated, when and how long. Social distress is one of the important components of the syndrome. This requires appropriate treatment to improve the quality of life of these patients. Mild cases of TS do not require any treatment, unless they are associated with ADHD and OCB which may markedly impact on daily life.

The basic treatment of TS is represented by drugs acting on the dopaminergic system such as haloperidol and pimozide [13, 14]. This latter, similarly to other selective D2 receptor antagonists, frequently causes undesirable side effects, especially ECG alterations, which require close monitoring. Sulpiride, tiapride and the recently introduced atypical neuroleptics risperidone and olanzapine have also been employed in TS with fairly good results. Tetrabenazine is a powerful dopamine depleting agent [15] which has shown good results in TS but the drug is not yet available on the Italian market. Presently, no clinical or laboratory data exist which may indicate which type of neuroleptic should preferentially be employed.

The complexity of the pathogenetic systems involved in the disease is shown by the fact that also the opposite pharmacological approach with dopamine agonists such as pergolide has been employed [16] while alpha-2 adrenergic agonists such as clonidine and guanfacine have also been used [11, 17].

CNS stimulating drugs such as methylphenidate and amphetamines have been shown to be useful in disturbances related to ADHD while selective serotonin re-uptake inhibitors (SSRI) have proven to be beneficial in OCB [18–22]. Recently, the transdermal administration of nicotine has been shown to be useful in reducing the dosage of neuroleptics [23].

Botulinum toxin, by local injection, has recently been successfully employed to relax muscles involved in tics: the muscle relaxation also reduces propriocep-tive activity, thus decreasing sensory feeling which is often the cause of the pathological motor fits (pre-monitory sensation). The percutaneous injection into vocal cords has been shown to be effective in reliev-ing muscle tension thus decreasing coprolalia and vocalizations. A transient hypophonia occurred in 85–90% of cases [24, 25].

Supportive psychiatric and psychodynamic treat-ments are currently used especially to improve sub-jective tolerability of the disturbance by patients and their family [26, 27].

Deep Brain Stimulation in Tourette Syndrome

Over the last few years, various reports [28, 29] have been published concerning surgical treatment of in-tractable Tourette Syndrome by means of stereotactic deep brain stimulation. On the bases of these prelimi-nary experiences, starting from November 2004 DBS was performed in 10 patients with chronic, severe to marked tics, resistant to at least 1 year of conservative treatment. Subjects were evaluated by a multi-disci-plinary team, and patients with cognitive impairment or psychosis were excluded. Patients have been treated targeting centromedian-parafascicular and ventral-oral complex nuclei of the thalamus, bilaterally. Coor-dinates are referred to the commissural line and are 5 mm lateral to the AC-PC line, 2 mm posterior to mid-commissural point, and at the AC-PC plane. Sur-gical procedure was performed under local anesthe-sia, with stereotactic guidance with the aid of neuro-navigation (Treon – Medtronic). The day before sur-gery an MRI scan of the brain was obtained: 3 mm thickness, with T1 weighed axial and sagittal, and T2 weighed coronal.

Preoperatively, a CT scan of the brain was ob-tained after positioning of the stereotactic frame un-der local anesthesia. The exam was then acquired by the neuronavigation device (TREON, Medtronic), and fused with the preoperative MRI, in order to obtain the coordinates of the nuclei according to the Schaltenbrand-Wahren atlas and the Talairach grid (◼ Figs. 19.1 and 19.2).

A pre-coronal drill hole of 14 mm diameter was then performed after proper local anesthesia. Using the "ben-gun" method, a three-channels simultaneous micro-registration was obtained with tungsten bi-polar micro-electrodes (Inomed) with high-imped-ance.

After a "map" of neuronal activity is thus drawn, we proceeded with macro-stimulation with a gradual increase of the intensity, in order to record every sub-jective impression reported by the patient. The limit of 5 m Amp was usually reached. Macro-stimulation with these features started at the target for all the three channels, and was performed for 2 minutes (accord-ing to clinical manifestations) at every millimeter along the stereotactic trajectory until reaching 5 mm above the target: the multipolar, definitive stimulating electrode was positioned following the trajectory of one of the three probes, considering the results of stimulation and the adverse effects.

Intralaminar and parafascicular thalamic nuclei and the internal part of the ventral-oral thalamic nucleus were targeted on the basis of the neuronavi-gation reports.

The day after the procedure a control T2 weighed, Inversion Recovery, 3 mm coronal MRI scan was ob-tained.

The two stimulating electrodes were connected with the Kinetra device (Medtronic), positioned in a subcutaneous, pre-pectoral fashion during a second surgical procedure performed under general anesthe-sia at a mean of 3 days after the first surgical proce-dure.

Results of surgical series are summarized in ◼ Table 19.1.

Patients were followed with monthly clinical eva-luation by a multidisciplinary team. Tic manifesta-tions, social functioning as reported by caregivers and by the patients themselves were carefully re-corded and fluctuations in psychological manifesta-tions were evaluated with standardized psychological scales (Yale-Brown Obsessive-Compulsive Scale, State Trait Anxiety Inventory, Symptoms Checklist 90-R) [30–32].

Yale Global Tic Severity Scale values for 8 out of the 10 patients treated with a minimum of 3 months follow-up as reported in Graph 1: a global decrease in frequency and severity of tics is indicated. Concern-ing drug intake after DBS in 2 pts, medication was stopped completely; in the others drug intake was re-duced to a range of 25 to 50 percent.

Conclusions and Future Directions

The TS has been revisited in the last few years and much in its psychopathological picture has changed. Epidemiology has shown that TS is an important dis-ease both for its prevalence and its social impact and

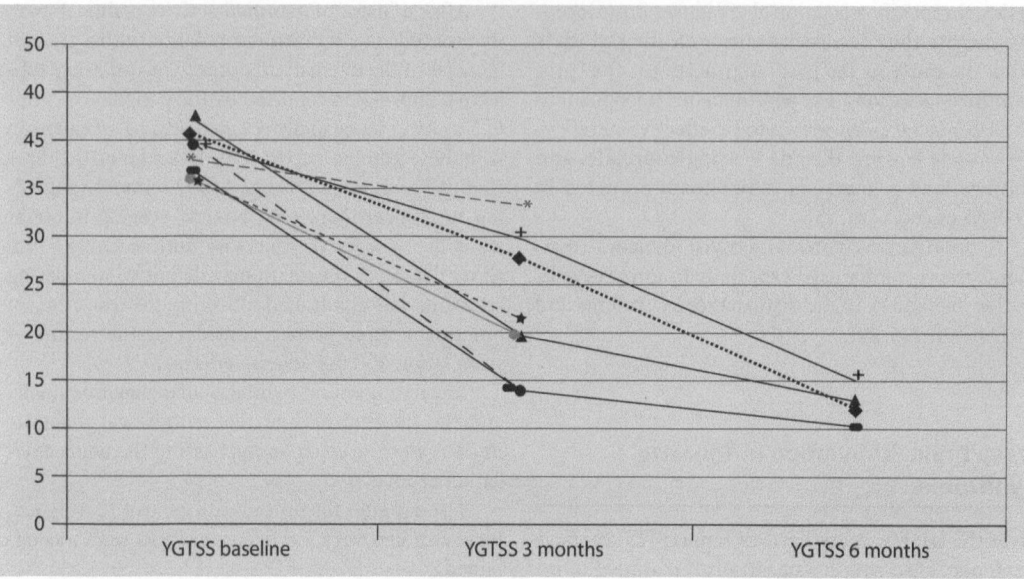

Fig. 19.1. Results of the YGTSS at the three follow-up endpoints.

◻ Table 19.1. Demographics and late follow-up of the surgical series

Name (abbr.)	Age	Symptoms/signs	Working status	Date of surgery	Stay (days)	Complications	Follow-up (January 2006)
CR	29	Complex motor, simple vocal, OCB	Student	November 2004	5	None	Minimal residual tics
FB	26	Complex motor, SIB, depression	Student	December 2004	6	None	Minimal residual tics
LT	56	Complex motor, complex vocal, depression	Unemployed	December 2004	8	None, thrombophlebitis four months after	Minimal residual tics
AL	45	Simple vocal, simple motor depression	Hardwork	January 2005	7	Psychomotor agitation	No tics
JS	19	Simple motor, complex vocal, palilalia, echolalia	Student	March 2005	8	None	Decrease in frequency of tics
SB	16	Complex motor, simple phonic, continuous manifestations	Student	March 2005	9	None	Minimal residual tics
SM	29	Complex motor, complex phonic, depression, OCB, anxiety	Sedentary work	April 2005	6	None	Minimal residual tics, OCB improved
SL	34	Complex motor, complex phonic, depression, anxiety, anti-social behavior	Hardwork	May 2005	6	None	Minimal residual tics, less depressed
FA	33	Complex motor, complex phonic, copro-echopalilalia, anti-social behavior	Unemployed	July 2005	3	Wound diastasis	Minimal residual tics
RB	33	Complex motor, complex phonic, OCB	Employee	July 2005	4	None	No residual tics, OCB improved

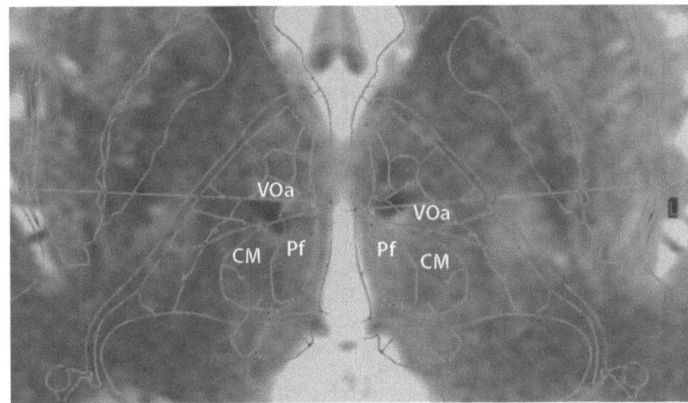

Fig. 19.2. Axial T2-IR, 3 mm sliced post-operative MRI, obtained 2 mm above the mid-commissural plane. The Schaltenbrand-Wahren Atlas has been superimposed to the MRI scan. *VO* Nucleus ventrooralis, *CM* Nucleus centralis magnocellularis, *Pf* Nucleus parafascicularis

that therefore it can no longer be considered as an orphan disease, while appropriate guidelines are being developed both for diagnosis and treatment.

An integrated, collaborative approach to treatment is mandatory, with the participation of neurologist, functional neurosurgeon, psychologist, psychiatrist and social operator.

This will help in improving quality of life of an important number of subjects who often find themselves in a situation of marked distress and significant social impairment.

References

1. American Psychiatric Association. Diagnostic and Statistical Manual of Mental Disorders. 4th edition, text revised (DSM-IV-TR). Washington DC, APA, 2000

2. Jankovic J. Phenomenology and classification of tics. Neurol Clin 1997;15: 267–275

3. Jankovic J. Diagnosis and classification of tics and Tourette's Syndrome. In: Chase T, Friedhoff A, Cohen DJ (eds) Tourette's Syndrome. Advances in Neurology. Vol. 58, New York Raven Press, 1992; pp 7–14

4. Robertson MM. Tourette Syndrome. Psychiatry 2005; 4: 92–98

5. Robertson MM, Trimble MR, Lees AJ. The psychopathology of the Gilles de la Tourette Syndrome: a phenomenological analysis. Br J Psychiatry, 1988: 152: 383–390

6. Paterson AI, Azrin NH. An evaluation of behavioral treatments for Tourette syndrome. Behav Res Therapy 1992; 30: 167–174

7. March JS. Cognitive-behavioral psychotherapy for children and adolescents with OCD: a review and recommendations for treatment. Am Acad Child Adolesc Psychiatry 1995; 34: 7–18

8. Peterson BS. Neuroimaging studies of Tourette syndrome: a decade of progress. In: Cohen DJ, Jankovic J, Goetz CG (eds) Tourette syndrome. Vol. 85, Advances in Neurology. Lippincolt Williams and Wilkins and Wilkins, Philadelphia; 2001, pp 179–96

9. Eidelberg D, Moeller JR, Antonini A, et al. The metabolic anatomy of Tourette syndrome. Neurology 1997; 48: 927–34

10. Leckman JF, Dolansky ES, Hardin MT. Perinatal factors in the expression of Tourette's syndrome: an exploratory study. J Am Acad Child Adol Psychiatry 1990; 29: 220–226

11. Cohn LM, Caliendo GC. Guanfacine use in children with attention deficit hyperactivity disorder (letter) Ann Pharmacol 1997; 31: 918–919

12. Pinelli P, Colombo R, Onorato S. Advances in Rehabilitation: Analisi dell'attenzione protratta nelle reazioni verbali. Maugeri Foundation Books, 1999, Vol.1, n 6

13. Sallee FR, NesbittL, Jackson C, Sine L, Sethurman G. Relative efficacy of haloperidol and pimozide in children and adolescents. Am J Psychiatry 1997; 154: 1057–1062

14. Rickards H., Hartley N., Robertson MM. Seignot's paper on the treatment of Tourette's syndrome with haloperidol. Hist Psychiatry 1997; VIII: 433–436

15. Jankovic J, Beach J. Long-term effects of tetrabenazine in hyperkinetic movement disorders. Neurology 1997; 48: 358–362

16. Lipinski JF, Sallee FR, Jackson C, Sethuraman G. Dopamine agonist treatment of Tourette disorder in children: results of an open-label trial of pergolide. Mov Disord 1997; 12: 402–407

17. Chappel PB, Riddle MA, Scahill L, Lynch KA, Scuktz R, Arnsten A et al. Guanfacine treatment of comorbid attention-deficit hyperactivity disorder and Tourette's syndrome: preliminary clinical experience. J Am Acad Adolesc Child Psychiatry 1995; 34: 1140–1146

18. Castellanos FX, Giedd JN, Elia J, Marsh WL, Ritchie GF, Hamburger SD. Controlled stimulat treatment of ADHD and comorbid Tourette's syndrome: effects of stimulant and dose. J Am Acad Child Adolesc Psychiatry 1997; 36: 589–596

19. Abwender DA, Como PG, Kurlan R, Parry K, Fett KA, Cui L, et al. School problems in Tourette's syndrome. Arch Neurol 1996; 53: 509–511

20. De Groot CM, Yeates KO, Baker GB, Bornstein RA. Impaired neuropsychological functioning in Tourette's syndrome subjects with co-occuring obsessive-compulsive and attention deficit symptoms. J Neuropsychiatry Clin Neursci 1997; 9: 267–272

21. Nolan EE, Gadow KD. Children with ADHD and tic disorder and their classmates: behavioral normalization with methylphenidate. J Am Acad Child Adolesc Psychiatry 1997; 36: 597–604

22. Scahill L, Riddle MA, King RA, Hardin MT, Rasmussen A, Makuch RW, Leckman JF. Fluoxetine has no marked effect on tic symptoms in patients with Tourette's syndrome: a double-blind placebo-controlled study. J Am Acad Adolesc Child Psychiatry 1997; 7: 75–85

23. Dursun SM, Reveley MA. Differential effects of transdermal nicotine on microstructured analyses of tic in Tourette's syndrome: an open study. Psychological Med 1997; 27: 483–487

24. Porta M, Maggioni GR, Ottaviani F, Schindler A. Treatment of phonic tics in patients with Tourette's Syndrome using botulinum toxin type A. Neurol Sci 2003; 24: 420–423

25. Scott BI, Jankovic J, Donovan DT: Botulinum toxin injection into vocal cord in the treatment of malignant coprolalia associated with Tourette's Syndrome. Mov Disord. 1996; 11: 431–3

26. Babel TB, Warnke PC, Ostertag CB. Immediate and long term outcome after infrathalamic and thalamic lesioning for intractable Tourette's syndrome. J Neurol Neurosurg Psychiatry 2001; 70: 666–671

27. Rausch S, Baer L, Cosgrove GR, Jenike MA. Neurosurgical treatment of Tourette's syndrome: a critical review. Compr Psychiatry 1995; 36:141–156

28. Visser-Vanderwalle V, Temel Y, Boon P, Freeling F, Colle H, Hoogland G, Groenewegen H, Van der Linden C. Chronic bilateral thalamic stimulation: a new therapeutic approach in intractable Tourette syndrome. Report of three cases. J Neurosurg 2003; 99: 1094–1100

29. Houeto JL, Karachi C, Mallet L et al. Tourette's syndrome and deep brain stimulation. J Neurol Neurosurg Psychiatry 2005; 76: 992–995

30. Goodman WK, Price LH, Rasmussen SA. The Yale-Brown Obsessive Compulsive Scale. I. Development, use, and reliability. Arch Gen Psychiatry 1989; 46 (11): 1006–1011

31. Bertolotti G, Michelin P, Sanavio E, Simonetti G, Vidotto G, Zotti AM. CBA 2,0 Cognitive Behavioral Assessment, 4th edn. Organizzazioni Speciali. Firenze, 1987

32. Derogatis LR. SCL 90-R, administration, scoring and procedures manual-I for the revised version. Baltimore: John-Hopkins University School of Medicine; 1977

Biographies of the Authors

Prof. Alberto Albanese

Alberto Albanese is Professor of Neurology at the Istituto Nazionale Neurologico Carlo Besta, Universita Cattolica del Sacro Cuore Milan, Italy. His primary research interest is in the use of deep brain stimulation in Parkinson's Disease and dystonia. He has published more than 190 papers. Professor Albanese is the Chair of the European Dystonia Study Group and a member of the advisory boards for the Neurotoxin Institute, the European Dystonia Federation and the European Parkinson's Disease Association.

Prof. Alim Luis Benabid

Alim Louis Benabid is Professor of Biophysics and Director of the Laboratory of Preclinical Neurosciences at INSERM, and Head of Stereotactic and Functional Neurosurgery at the Joseph Fourier University Hospital, Grenoble, France. In addition, he is the Coordinator of the Claudio Munari Center for Functional Neurosurgery, Milan. He received his MD and his PhD in Physics at the Joseph Fourier University of Grenoble (UJFG), and completed his basic neurosciences training at the Salk Institute California. He was responsible for the first pilot study for the use of deep brain stimulation to treat movement disorders and his current research explores the mechanism of action and the acute and long-term biological effects of this technique.

Prof. Benabid is an elected member of the French Academy of Medicine, the Royal Academy of Medicine in Belgium and the Academy of Sciences at the Institut de France. He has published over 200 papers and several book chapters. He is on the editorial board of *Stereotactic and Functional Surgery*, *Movement Disorders* and *Neurosurgery*.

Prof. Paul Boon

Paul Boon is Professor of Neurology at Ghent University Hospital, Belgium and chairman of the Research and Development Committee and coordinator of the MRI research group at Kempenhaege, a large epilepsy and sleep disorders clinic. He graduated as an MD from Ghent University before further training at the University of Texas and the Bowman-Gray Medical School, North Carolina. His main research interests are clinical epilepsy, quantitative EEG analysis, source localization and anticipation, neuromodulation and functional neuroimaging. He has published over 150 papers in the fields of epilepsy, neuropsychiatry and sleep. He is a member of the executive committee of the European Neurological Society, board member of the Belgian Neurological Society and a member of several commissions of the International League against Epilepsy and the American Epilepsy Society. He is editor-in-chief of Acta Neurologica Belgica.

Prof. Giovanni Broggi

Giovanni Broggi is Professor and Director of Neuro-surgery at the Istituto Nazionale Neurologico Carlo Besta, in Milan, Italy. He studied medicine and surgery at the University of Milan, trained in neurology first at the University of Parma and then at the University of Milan. His main research interest is in the use of deep brain stimulation for the treatment of movement disorders, epilepsy and chronic pain. He has published over 250 papers and 10 books in the field of neurosurgery. He is the Vice-President of the ESSFN (European Society for Stereotactic and Functional Neurosurgery) and is a member of the editorial board of *Neurosurgery*.

Professor Broggi is also an active member of the following Societies: SINCH, (Società Italiana di Neurochirurgia), SIN (Società Italiana di Neurologia, of which he was past President), AANS (American Association of Neurological Surgeons) CNS (College of Neurological Surgeons). Past President of SINCH.

Prof. Eric Buchser

Eric Buchser is Professor of Anaesthesia at the University Hospital of Lausanne, Switzerland and Head of the Anaesthesia and Pain Management Services at the Morges Hospital, Switzerland. He is the chairman of the oral examination of the European Diploma of Anaesthesia and Intensive Care of the European Society of Anaesthesiology. He has been involved in cooperation programmes on anaesthesia and pain management in Vietnam and Tanzania for more than 10 years. One of his main research interests is the evaluation of physical activity in patients treated with spinal cord stimulation for chronic pain. He has published more than 30 papers.

Prof. Paul Cosyns

Paul Cosyns is Professor of Psychiatry at the University of Antwerp, Belgium. His current research interests are in the treatment of therapy refractory anxiety and depressive disorders: deep brain stimulation for the treatment of obsessive compulsive and depressive disorders, genetic factors in mood disorders and the role of hypothalamic-pituitary-adrenal axis dysfunction in mood disorders. He is President of the Belgian College of Neuropsychopharmacology and Biological Psychiatry, a member of the board of the European Association of Psychiatrists and of the Belgium consultative committee of Bio-ethics. He is also involved in forensic psychiatry research (treatment of sexual abusers).

Prof. Giorgio Cruccu

Giorgio Cruccu is Professor of Neurology at La Sapienza University, Rome. His main research field is clinical neurophysiology of pain, with a particular interest for the trigeminal system: He has published over 130 articles and 30 book chapters. He is chairman of the Panel on Neuropathic Pain of the European Federation of Neurological Sciences (EFNS), co-ordinator of the Italian Study Group on Neuroscience and Pain, and president of the Brain Stem Society.

Prof. Marshall Devor

Prof. Marshall Devor is Chairman of the Department of Cell and Animal Biology at the Institute of Life Sciences of the Hebrew University of Jerusalem. He received his BA degree from Princeton University, and his PhD from the Department of Brain and Cognitive Sciences at the Massachusetts Institute of Technology in Cambridge, Massachusetts.

Through his research, Prof. Devor has contributed considerably to the understanding of the neurobiologic basis of neuropathic pain. His laboratory has published extensively in the pain field, with work of a notably integrative nature involving neurophysiology, computer simulations, neuroanatomy (light and electron microscopy), genetics, and animal behavior. He is the author of well over 200 research publications in the field of pain science.

He has served as a member or chair of a variety of professional committees of IASP, EFIC, and of the World Institute of Pain (WIP). He is a Section Editor of Pain, and a member of the Editorial Boards of many journals, including European Journal of Pain, Journal of Basic and Clinical Physiology and Pharmacology, The Pain Clinic, Pain Practice, Current Pain and Headache Reports, and Journal of Neuropathic Pain & Symptom Palliation. Prof. Devor is also a Section Editor of the Encyclopedic Reference of Pain.

Angelo Franzini

Angelo Franzini is a neurosurgeon in the Department of Neurosurgery at the Neurologic Institute Carlo Besta in Milan, Italy, and has contributed greatly to the creation of the Functional Neurosurgery division within the department. Dr Angelo is a specialized stereotactic neurosurgeon and has developed a digital atlas of the basal ganglias. He also has interests in microsurgery of peripheral nerves, back surgery, neurosurgery of tumors and movement disorders in general, especially the treatment of Parkinson's Disease.

Dr. Loes Gabriëls

Loes Gabriëls is a psychiatrist working at the University Hospital in Antwerp, Belgium. She graduated from the University of Gent with a Masters degree in Engineering, followed by a medical degree from the University of Antwerp. Dr Gabriëls has special interests in obsessive-compulsive disorder (OCD), anxiety and sleep problems, and is involved in a research project focusing on the pathophysiology and treatment of OCD. Neurobiological models propose that hyperactivity within circuits linking the cerebral cortex, basal ganglia and thalamus is crucial to OCD symptoms. Dr Gabriëls has studied the therapeutic effects of electrical stimulation on these circuits in patients with severe and treatment-refractory illness. Dr Gabriëls is also Treasurer of the Benelux Neuromodulation Society.

Prof. Peter Goadsby

Peter Goadsby is Professor of Clinical Neurology, Institute of Neurology, University College London and Honorary Consultant Neurologist at the National Hospital for Neurology and Neurosurgery, Queen Square, and at the Hospital for Sick Children, Great Ormond St, London. He leads the Headache Research Group at Queen Square. He studied medicine in Sydney and trained in neurology under Prof James Lance. His major research interests are in the neural control of the cerebral circulation and the basic mechanisms of head pain in both experimental settings and in the clinical context of headache. He has published over 300 papers in the fields of pain and neurology, is Editor-in-Chief of Cephalalgia, and on the editorial boards of numerous neurology journals. In addition, he is a medical trustee of the Migraine Trust.

Prof. Bengt Linderoth

Bengt Linderoth is Professor of Neurosurgery at the Karolinska Institutet and head of the Section of Functional Neurosurgery at the Karolinska University Hospital, Stockholm, Sweden. Dr Linderoth has also directed the Functional Neurosurgery Research Group at the Karolinska for more than seven years both concerning experimental animal projects and

clinical studies. His main research interest is in the physiological mechanisms underlying neuropathic and vasculopathic pain and the beneficial effects of electric CNS stimulation, especially spinal cord stimulation in a number of disorders, particularly neuropathic pain after nerve injury, peripheral ischemic conditions and lately also cardiac ischemia with angina pectoris.

He has published more than 70 papers and written many reviews and chapters in the field of neuromodulation. He is a member of the executive committee of the European Society for Stereotactic and Functional Neurosurgery; board of directors in the World Society (WSSFN) and is a member of the editorial board of Neuromodulation.

Prof. Jean-Paul Nguyen

Jean-Paul Nguyen is Professor of Neurosurgery at Henri Mondor University Hospital, Creteil, France. He studied medicine and then neuroanatomy at the Faculty of Medicine, Paris. His main research interests are stereotaxy, functional neurosurgery, intraputaminal grafts and deep brain stimulation for the treatment of movement disorders, and neurosurgical treatment of chronic pain. He has published over 80 articles on these subjects. He is a member of the editorial board of Neurosurgery and on the scientific board of the French Society of Neurosurgery.

Prof. Bart Nuttin

Bart Nuttin is Professor of Neurosurgery at the Katholieke Universiteit Leuven, Belgium, and is responsible for stereotactic and functional neurosurgery in the Universitaire Ziekenhuizen Leuven. He received his MD and PhD in Leuven, but trained additionally in Uniklinikum Köln with Prof. V. Sturm and in Brigham and Women's and Children's Hospital in Boston with Prof P. Black. His group published the first pilot study on electrical stimulation of the anterior limbs of the internal capsule and the grey matter below for treatment-refractory obsessive-compulsive disorder. Together with the DBS-OCD collaborative group he established guidelines for electrical brain stimulation for psychiatric disorders. His main research interest is still DBS for psychiatric disorders in humans. In the lab his group studies electrical brain stimulation (indications, mechanisms, devices) in different animal models for psychiatric disorders. He was appointed to the council of the "Europäische Akademie zur Erforschung von Folgen wissenschaftlich-technischer Entwicklungen Bad

Neuenahr-Ahrweiler GmbH", he is the president of the Benelux Chapter of the International Neuromodulation Society and is a member of the Executive Committee of the European Society for Stereotactic and Functional Neurosurgery.

Prof. Mauro Porta

Mauro Porta is specialized in neurology and neurosurgery. He is Director of the Department of Neurology with Tourette Syndrome Center and Movement Disorders Clinic, at Policlinico San Marco, Zingonia-Bergamo, Italy. The department has also a Pain and Headache Center involved in advanced treatment procedures. Mauro Porta is consultant at Istituto Galeazzi, Milan, Italy, where Dr. D. Servello, functional neurosurgeon, has his clinical activity. His main interests are Tourette Syndrome, botulinum toxin and pain. He is leading a study on Deep Brain Stimulation in Tourette Syndrome. He has published over 150 papers and has edited a number of books on neuroscience. He is President of the Italian Assosiation for Tourette Syndrome.

Prof. Thomas Schlaepfer

Thomas Schlaepfer is Vice Chair of Psychiatry and Psychotherapy at the University of Bonn, Germany; he is jointly appointed as Associate Professor of Psychiatry and Mental Hygiene at the Johns Hopkins Hospital. He studied medicine at the University of Bern Switzerland, and holds an Executive Medical-Business Graduate Certificate from John Hopkins University. His primary research interests are the use of different brain stimulation methods as putative treatments of affective disorders, cytokine induced depression and functional neuroimaging in affective and substance abuse disorders, on which he has published over 80 papers. He is on the editorial boards and acts as a reviewer for numerous CNS journals and is a member of the American College of Neuropsychopharmacology (ACNP).

Dr. Philip Starr

Philip Starr is Associate Professor of Neurological Surgery, Co-Director of the Functional Neurosurgery Program and Principal Investigator in the Movement Disorders Research division at the University of California, San Francisco, USA. He also holds the post of Surgical Director of the Parkinson's Disease Research Education and Care Centre (PADRECC) at San Francisco Veterans Affairs Medical Centre. He studied for

his first degree at Princeton University, before obtaining his MD and PhD at Harvard Medical School. In addition, he has fellowship training in micro-electrode guided surgery for movement disorders, which he completed at Emory University in Atlanta. His research is focused on the neurophysiology and therapy of movement disorders and specific interests include physiology of the basal ganglia, clinical trials of novel surgical therapies in movement disorders and the use of interventional magnetic resonance imaging (iMRI). He has published more than 25 papers on these subjects, including 11 on deep brain stimulation.

Prof. Volker Sturm

Volker Sturm is Professor, Chairman and Director of Stereotactic and Functional Neurosurgery at the University of Cologne, Germany. He completed his training in medicine and neurosurgery at the University of Heidelberg. His major research interests are functional neuroimaging and deep brain stimulation. He has published more than 125 papers on these subjects. He is an honorary member of the German Society of Medical Physics and a member of the Functional Neurosurgery. He has recently been awarded the Erwin Schroedinger Prize for his development of a brain pacemaker.

Prof. Rod Taylor

Rod Taylor is a Reader in Public Health and Epidemiology and Director of the Masters program in Health Technology Assessment at the University of Birmingham, UK. From 1999 to 2001, he was first Director of Appraisals at National Institute for Health and Clinical Excellence (NICE) and is currently a member of one of its three appraisal committees. His research interests include the use of evidence in healthcare policy-making and methods of evidence synthesis and decision analysis. He has published over 70 papers. He is a biostatistics and health economics reviewer to a number of peer review journals and serves on a number of UK government and academic advisory panels.

Rod has a Masters in Medical Statistics, from the London School of Hygiene Tropical Medicine at the University of London. He also has a Postgraduate Certificate in Health Economics and Health.

Dr. Chris Van der Linden

Chris van der Linden is a Doctor Neurologist in movement disorders and neurology, and head of the Movement Disorders Centre at the Sint-Lucas Hospital, Ghent, Belgium. He trained at the University of Texas Medical School and the Baylor College of Medicine, Texas. His major research interest is in the use of deep brain stimulation for the treatment of Parkinson's disease and Tourette's disorder and he has published more than 25 papers in the field of movement disorders. He is a member of the Medical Advisory Board of the European Parkinson's Disease Association.

Dr. Bob Van Hilten

Bob van Hilten is Associate Professor of Neurology at Leiden University Medical Centre, the Netherlands. His research is focused on the phenotyping and the longitudinal course of Parkinson's disease and the movement disorders of complex regional pain syndrome. He has published over 90 papers in the field of movement disorders including work on the use of intrathecal baclofen therapy in dystonia of complex regional pain syndrome. He is a member of the European executive committee of the Movement Disorder Society and scientific director of the Trauma Related Neuronal Dysfunction (TREND) consortium.

Dr. Veerle Visser-Vandewalle

Veerle Visser-Vandewalle studied medicine at the State University in Ghent in Belgium and was trained as a neurosurgeon in the Sint-Jans Hospital in Brugge (Belgium). In 1999, she became member of the staff in the Department of Neurosurgery at the University Hospital of Maastricht in the Netherlands. Her work has been focussed on Functional Neurosurgery, more specifically on deep brain stimulation. In 2004 Dr. Visser-VandeWalle gained her Ph.D on the subject: "Deep Brain Stimulation in Movement Disorders: Applications reconsidered".